T0301865

Advanced Mobile Technologies for Secure Transaction Processing:

Emerging Research and Opportunities

Raghvendra Kumar
LNCT Group of College, India

Preeta Sharan
The Oxford College of Engineering, India

Aruna Devi
Surabhi Software, India

A volume in the Advances in Wireless
Technologies and Telecommunication
(AWTT) Book Series

Published in the United States of America by
    IGI Global
    Information Science Reference (an imprint of IGI Global)
    701 E. Chocolate Avenue
    Hershey PA, USA 17033
    Tel: 717-533-8845
    Fax: 717-533-8661
    E-mail: cust@igi-global.com
    Web site: http://www.igi-global.com

Library of Congress Cataloging-in-Publication Data

Names: Kumar, Raghvendra, 1987- author. | Sharan, Preeta, 1967- author. |
  Devi, Aruna, 1964- author.
Title: Advanced mobile technologies for secure transaction processing :
  emerging research and opportunities / by Raghvendra Kumar, Preeta Sharan,
  and Aruna Devi.
Description: Hershey, PA : Information Science Reference, 2017. | Includes
  bibliographical references.
Identifiers: LCCN 2017010706| ISBN 9781522527596 (hardcover) | ISBN
  9781522527602 (ebook)
Subjects: LCSH: Mobile commerce--Security measures. | Electronic
  commerce--Security measures.
Classification: LCC HF5548.32 .K858 2017 | DDC 332.1/78--dc23 LC record available at https://
lccn.loc.gov/2017010706

This book is published in the IGI Global book series Advances in Wireless Technologies and
Telecommunication (AWTT) (ISSN: 2327-3305; eISSN: 2327-3313)

Advances in Wireless Technologies and Telecommunication (AWTT) Book Series

ISSN:2327-3305
EISSN:2327-3313

Editor-in-Chief: Xiaoge Xu, Xiamen University Malaysia, Malaysia

MISSION

The wireless computing industry is constantly evolving, redesigning the ways in which individuals share information. Wireless technology and telecommunication remain one of the most important technologies in business organizations. The utilization of these technologies has enhanced business efficiency by enabling dynamic resources in all aspects of society.

The **Advances in Wireless Technologies and Telecommunication Book Series** aims to provide researchers and academic communities with quality research on the concepts and developments in the wireless technology fields. Developers, engineers, students, research strategists, and IT managers will find this series useful to gain insight into next generation wireless technologies and telecommunication.

COVERAGE

- Wireless sensor networks
- Broadcasting
- Global Telecommunications
- Mobile Communications
- Network Management
- Wireless Broadband
- Grid Communications
- Digital Communication
- Cellular Networks
- Wireless Technologies

IGI Global is currently accepting manuscripts for publication within this series. To submit a proposal for a volume in this series, please contact our Acquisition Editors at Acquisitions@igi-global.com or visit: http://www.igi-global.com/publish/.

Titles in this Series

For a list of additional titles in this series, please visit:*
https://www.igi-global.com/book-series/advances-wireless-technologies-telecommunication/73684

Routing Protocols and Architectural Solutions for Optimal Wireless Networks and Security
Dharm Singh (Namibia University of Science and Technology, Namibia)
Information Science Reference • © 2017 • 277pp • H/C (ISBN: 9781522523420) • US $205.00

Sliding Mode in Intellectual Control and Communication...
Vardan Mkrttchian (HHH University, Australia) and Ekaterina Aleshina (Penza State University, Russia)
Information Science Reference • © 2017 • 128pp • H/C (ISBN: 9781522522928) • US $140.00

Emerging Trends and Applications of the Internet of Things
Petar Kocovic (Union – Nikola Tesla University, Serbia) Reinhold Behringer (Leeds Beckett University, UK) Muthu Ramachandran (Leeds Beckett University, UK) and Radomir Mihajlovic (New York Institute of Technology, USA)
Information Science Reference • © 2017 • 330pp • H/C (ISBN: 9781522524373) • US $195.00

Resource Allocation in Next-Generation Broadband Wireless Access Networks
Chetna Singhal (Indian Institute of Technology Kharagpur, India) and Swades De (Indian Institute of Technology Delhi, India)
Information Science Reference • © 2017 • 334pp • H/C (ISBN: 9781522520238) • US $190.00

Multimedia Services and Applications in Mission Critical Communication Systems
Khalid Al-Begain (University of South Wales, UK) and Ashraf Ali (The Hashemite University, Jordan & University of South Wales, UK)
Information Science Reference • © 2017 • 331pp • H/C (ISBN: 9781522521136) • US $200.00

Big Data Applications in the Telecommunications Industry
Ye Ouyang (Verizon Wirless, USA) and Mantian Hu (Chinese University of Hong Kong, China)
Information Science Reference • © 2017 • 216pp • H/C (ISBN: 9781522517504) • US $145.00

For an enitre list of titles in this series, please visit:
https://www.igi-global.com/book-series/advances-wireless-technologies-telecommunication/73684

701 East Chocolate Avenue, Hershey, PA 17033, USA
Tel: 717-533-8845 x100 • Fax: 717-533-8661
E-Mail: cust@igi-global.com • www.igi-global.com

Table of Contents

Preface

The rapid technological advancements in cellular communications, wireless LAN and satellite services have led to the emergence of mobile computing. In mobile computing, users are not attached to a fixed geographical location; instead their point of attachment to the network changes as they move. The emergence of a computing platforms such as smart phones, laptops and personal digital assistants (PDA) have made possible for people to work from anywhere at any time via wireless communication network. With the advancement in technology millions of users carry portable computers and communication devices use a wireless connection to access worldwide global network. Each mobile unit equipped with wireless network can be connected to global information network to provide unrestricted user mobility. Mobility and portability pose new challenges to the mobile database management and distributed computing. The database software support for mobile computing is still in the germinating stage. There is necessity to design specifications for distributed methodologies across the desperate system. This is required to develop database software systems that extend existing database systems designs and platforms to satisfy the constraints imposed by mobile computing. One of the major challenges is how to handle long period of disconnection, other constrained resources of mobile computing such as limited battery life and variable bandwidth etc. In mobile computing, there will be more competition for shared data since it provides users with ability to access information and services through wireless connections that can be retained even while the user is moving. Further, mobile users will have to share their data with others. The task of ensuring consistency of shared data becomes more difficult in mobile computing because of limitations and restrictions of wireless communication channels.

It is very important to understand how mobile computing differs from distributed database computing. How does mobility affect transaction processing and replication. Whether the location management can be the integral part of the database management problem. Further is it essential to know how to replicate the location data. Caching is key to handle frequent disconnection in mobile computing. The challenges are to keep cache consistent with minimum communication cost. Also it is important to know how query processing is different in mobile computing environment.

The mobile computing paradigm has emerged due to advances in wireless networking technology and portable computing devices. Mobile computing enables users equipped with portable computing devices to access information services through a shared infrastructure, regardless of physical location or movement. The mobile computing environment is a distributed computing platform with the following differences: the mobility of users and the mobile devices access the limited computational and power sources.

Mobile users now have the ability to send and retrieve emails, receive updates on stock prices and weather, and obtain driving directions while in motion using cellular phones, pagers, and PDAs. Wireless transmission media across wide-area network. Transaction Processing Model for the Mobile Data Access System communication networks are also an important element in the technological infrastructure of E-commerce. The effective development of guided and wireless media networks will support the delivery of World Wide Web functionality over the Internet. Using mobile technologies will enable users to purchase E-commerce goods and services anywhere and anytime. Naturally, mobile users also desire the same functionality available to them at a stationary computer on a wired network and save changes to documents stored on a file server or to query and update shared data in private or corporate databases.

As mobile users wander about, they are bound to encounter a variety of different information sources (databases) that are often autonomous and heterogeneous in nature. It would be advantageous if a uniform interface can be presented to the mobile users freeing them from the need to have knowledge of the data representation or data access method employed at different data sources. Organizing a collection of autonomous databases into a multi database is therefore desirable. A multi database integrates pre-existing autonomous and heterogeneous databases to form a global distributed information-sharing paradigm. To support mobile users, it is necessary to augment the existing multi databases with wireless networking capabilities. This augmented multi database is known as a Mobile Data Access System (MDAS).

Mobile database replication technology is an effective way to support disconnected operation, with limitations. Due to the need to save huge information's on a mobile device, when the data size increases the limited resources of mobile devices limit the overall size of the information storage capability. In addition of replicated mobile database systems require periodic synchronization to ensure the consistency of the database system, so for users to access popular data only occasionally in this way should not be processed. A transaction is an important database system concept. In a data base system, the user application to the transaction is the basic unit to achieve access to the database. A transaction is composed of a series of read write operations, these operations begin with transactions to abort or commit. In order to improve the efficiency of the database, each transaction can be executed concurrently, to ensure the serializability of transaction scheduling by the concurrency control mechanism of the database. Serializability of transaction scheduling a database system is to determine the correct standard. This book discusses some of the problems identified with mobile database computing and explores the upcoming research challenges. Some of the problems involved in supporting transaction services and distributed data management in a mobile environment has been identified. The book is organized into five chapters:

Chapter 1 discusses the factors that affect trust and its development vary from traditional banking services because of the uncertain nature of the online environment. Extensive efforts in identifying factors that affect trust have shown security to play an important role in its development. Every step in the online banking activities of users are secured by one of more security mechanisms. Analyzing the role of these mechanisms in developing a user's perception of security and the impact of this perception on trust provides a pathway to study the role of security in trust development. When designing a security protocol, it is important firstly to define its security goals. These goals are generally referred to as security services. It is also important to identify the mechanisms which can provide each security service. In this chapter, therefore, define the security services of relevance to this chapter and after that the security mechanisms, which can be used to provide the security services are then examined.

Chapter 2 discussed different security aspects for mobile banking in the context of mobile devices and applications, a few common themes emerge. If these themes are not addressed properly, through security controls and measures, the underlying threats could compromise the confidentiality, integrity and availability of mobile security assets. Mobile security assets that need protection are mobile devices, the mobile application and private information. It is important that any new product or service is based on user needs and requirements rather than being driven by technology. This is an important issue to consider with SMS and mobile banking. Research related to the usability design issues surrounding online banking are relevant to SMS banking. Concepts related to authentication, online banking processes, and other relevant information security issues are discussed further. Literature relevant to these factors is also considered throughout this process of reflection. This chapter discussed the concepts related to authentication, online banking processes and other relevant information security issues are discussed.

Chapter 3 addresses about the incremental growth of mobile phones and mobile services even in poorer communities in the developing world (where the number of mobile phones has exceeded that of bank accounts) has led to an ever-larger number of services aimed at providing development in various sectors of the economies of the developing countries. Recent studies have shown the rise in the use of mobile applications to leapfrog the developmental agenda of many developing countries. This is because of continuous improvements in mobile technologies, increased affordability and availability. One of such mobile services developed to bring financial services to the rural unbanked is referred to as mobile money. Online transaction payment for the purpose of introducing web-based electronic money as an alternative way of online transaction payment, the main areas that cover in this chapter include research of current payment system, limitation of current payment, what are e-money and the current state of electronic money. It will discuss the proposed web-based electronic money as an alternative for online payment and the benefit of web-based e-money. Online payment transaction is a form of a financial exchange that takes place between the buyer and seller facilitated by means of electronic communications for conducting e-commerce and online purchasing. This chapter includes research of current payment system, limitation of current payment, what are e-money and the current state of electronic money.

Chapter 4 discussed about the concepts of Mobile banking. It is attractive because it allows people to do banking anytime, anywhere. One of the requirements of performing a mobile banking transaction is that users are required to login before use. In this chapter, we first revisit the basic concepts of database transactions, and discuss how these concepts are achieved in practical systems. Next, we briefly go through the architecture of transaction processing systems in the centralized and the distributed environments. This chapter we have reviewed the basic concepts of database systems and database transactions, and discussed the architecture of transaction processing systems in distributed environments. We will shift our focus to transactions and transaction processing in mobile environments, which possess some unique characteristics such as the mobility of mobile computing hosts, the limitations of wireless communications and the resource constraints of mobile computing devices.

Finally, Chapter 5 gives the concepts of the recent information technology development has vastly helped in accelerating and facilitating the banking services and operations in general. In spite of this accelerated development in the banking sector, the risk of invading electronic banking systems is evident. This is manifested in many harmful functions such as unauthorized money transfer, disclosure of client information, denial of online banking services as well as various threats linked with online banking at different lineages especially through authentication of the client online. The Payment Proxy Server monitors and analyses the content headers in HTML and WML pages moving between the content providers and merchants. Whenever it intercepts priced content it initiates a payment transaction and redirects the user to the Payment Server. Pricing and provisioning of the digital content and the payment service can be done through the administration tools. These tools are also used to monitor the events occurring in the payment server. In this chapter, we describe a system that achieves a secure data transmission over a mobile voice channel with a further goal to provide secure voice transmission. Here we describe each component of the system in detail and discuss the issues that we encountered while building the system.

Raghvendra Kumar
LNCT College, India

Preeta Sharan
The Oxford College of Engineering, India

Aruna Devi
Surabhi Software, India

Acknowledgment

Raghvendra Kumar likes to acknowledge the most important persons of his life, his grandfather Shri. Om Prakash Agrawal, grandmother Smt. Sheela Agrawal, father Mr. B.P. Agrawal, mother Mrs. Anita Agrawal, and finally, thanks to his brother-in-law Mr. Anshu Agrawal and sister Mrs. Neha Agrawal, as well as his sweetheart Priynka Pandey. This book has been his long-cherished dream which would not have been turned into reality without the support and love of these amazing people. These people encored him despite his falling in guiding them in proper time and attention.

Preeta Sharan is thankful to her son Aditya Prakash for his patience throughout her work. Further all her dear friends and last not the least her family member for their support. There have been several influences from her family and friends who have sacrificed lot of their time and attention to ensure that we are kept motivated to complete this crucial project.

Aruna Devi is thankful to her family members for their support rendered throughout this special project. Her special thanks are to her son Rohit Gupta and daughter-in-law Vishu Karnwal for their endless ideas and feedback. Dr. Surabhi Gupta her daughter needs a special mention for her motivation, inspiration and patience with this project who always encouraged in publishing the book.

There have been several influences from our family and friends who have sacrificed lot of their time and attention to ensure that we are kept motivated to complete this crucial project.

The authors are thankful to all the members of IGI Global, especially Jordan Tepper and Jan Travers for the given opportunities to write this book.

Raghvendra Kumar
LNCT College, India

Preeta Sharan
The Oxford College of Engineering, India

Aruna Devi
Surabhi Software, India

Chapter 1
Security Issues and Requirements for Online Banking

ABSTRACT

When designing a security protocol, it is important firstly to define its security goals. These goals are generally referred to as security services. It is also important to identify the mechanisms which can provide each security service. In this chapter, therefore, we first defined the security services of relevance to this chapter and after that the security mechanisms, which can be used to provide the security services are then examined. In next chapter gives the brief details about Attacks and Authentication for Mobile Banking.

INTRODUCTION

Development of digital technologies has transformed the transaction business data through E-commerce software like e-banking, e-tailing and it can extended as electronic mediums along with the internet and mobile phone are incorporated into daily activities. Electronic banking (e-trade programs), gives a use of physical medium to mobile banking transactions, that e-banking accommodates (Tan, Sutherland 2004) of greater than 50% of all banking

DOI: 10.4018/978-1-5225-2759-6.ch001

transaction. These consist of transactions over all electronic mediums along with ATMs, telephones and the internet. The internet particularly has innovative e-commerce. Internet banking has been gaining recognition is the recent years as it affords an on hand, faster and a minimum-value technique for people to perform their banking supports through an internet interface. That supports a person can perform via e-banking include retrieving account information such as transaction history and stability, shifting cash among identical consumer's bills or to exceptional consumer money, paying bills, and so forth. But, the acceptance of the mobile as a device for banking has been hampered by using several concerns. One of the largest challenges facing the enterprise, like e-trade (Yousafzai, Pallister, Foxall, 2005) of the customers to discover, that agree with is a crucial components to accelerate the growth of on-line programs. The internet primarily based on transactions which are characterized by uncertainty, anonymity and presents possibilities. The uncertain and unreliable nature of the underlying community could make consider an important element in this online environment. Perception of risk within the on line surroundings has a tendency to be better than most different electronic systems together with the online environment.

Understanding the character of consumer requirements and the elements, which affect crucial clues for its improvements in online e-banking, that are associated with the website, there are several features, which includes design, navigation, content, security, usability and so forth. Security mechanisms utilized by websites are mentioned as one of the maximum crucial functions that influence consumer needs. The banking scenario is incredibly dependent on security for the consumer. Thus, in digital banking, the function of protection and privateers will become critical for consumers trust. Security of client and server information is some other fundamental mission dealing with the digital trade enterprise, which are safety issues and concerns are group with the clients, capability or different elements are controlled by virtual services. Security is frequently as being the single important situation for client and server in e-trade and its associated components. The recognition of security technique for banking is often controlled by using the issues of both cutting-edge and capacity customers. The network is taken into consideration to be an unsecured medium and packages constructed. The use of net-banking for trade has resulted in an open for flexible community over the variety of functions, which may supplied to customers; however this has additionally multiplied the vulnerabilities that could result in decreased protection in such online systems. Internet banking is constructed on public networks might make greater at risk of security. Consumers have extra religion in different highly new networks such as mobile networks no matter comparatively minimum privacy (Accetta, Baron, Bolosky, Golub, Rashid, Tevanian, Young 1986). Resolving the issues of privacy and agree with in e-banking has been undertaken through the industry and the academia.

SECURITY IN E-BANKING

Security, being a complex concept, has been described by means of several researchers the usage of various category strategies. In widespread, safety is defined because the protection towards protection threats. In digital trade, a hazard could be defined as a spoil, alter, waste, deny or expose statistics or reduce performance of the facts and network components. These threats could appear on the server side and these could originate due to human, system or communication faults. Thus, the safety of the two susceptible factors in e-trade structures, which are the uncertain underlying technological infrastructure and the unreliable users of the gadget. Security is generally composed of a fixed of protection primitives or objectives. Each of those targets goals at defensive the systems and/or customers against threats. Over time, several different targets such as authentication and authorization have been diagnosed and been protected as a critical part of protection. The safety objectives of e-trade and its programs are confidentiality, integrity, availability, authentication, authorization, non-repudiation and privateers.

SECURITY SERVICES

There are five main security services (Accetta, Baron, Bolosky, Golub, Rashid, Tevanian, and Young 1986) which are of importance in this chapter. They are confidentiality, integrity, authentication, non-repudiation, and access control (Acharya, Franklin, and Zdonik 1995).

Confidentiality

Confidentiality means that the assets of a computer system and transmitted information and/or data are protected against access by unauthorized entities. Possible methods of access include printing, displaying, and other forms of disclosure, including simply revealing the existence of the information. According to ford, 'confidentiality services protect against information being disclosed or revealed to entities not authorized to have that information.'

Integrity

Integrity means that the assets of a computer system and transmitted information and/or data can be modified only by authorized parties and only in Data integrity services therefore are 'safeguards against the threat that the value or existence of data might be changed in a way inconsistent with the recognized security policy. Modification or changing the value of a data item includes writing, changing, changing the status, deleting, substituting, inserting, reordering, and delaying or replaying of transmitted messages. The Clark-Wilson model defines integrity as those qualities which give data and systems both internal consistency and a good correspondence to real- world expectations for the systems and data. Controls are needed for both internal consistency and external consistency.

Authentication

Authentication can be subdivided into origin authentication and entity authentication. Origin authentication provides corroboration to an entity that the source of received message is as claimed. The data origin authentication service provides the corroboration of the source of a data unit. However, the service in itself does not provide protection against duplication or modification of data units. Entity authentication ensures that an identity presented by a remote party participating in a communication connection or session is genuine. It is 'an ability to verify an entity's claimed identity, by another entity.

Non-Repudiation

Non-repudiation is the ability to prove that an action or event has taken place, so that this event or action cannot be repudiated later. In other words, the non repudiation service provides protection against one party to a communication

Non-repudiation of receipt or transmission provides the sender or the receiver respectively with the means to establish that the message was indeed received or transmitted. According to Ford, a non-repudiation service, in itself, does not eliminate repudiation. He states that 'it does not prevent any party from denying another party's claim that something occurred. What it does is ensure the availability of irrefutable evidence to support the speedy resolution of any such disagreement.

Access Control

The goal of an access control service is to protect against unauthorized access to any resource, for example a computing resource, communications resource, or information resource. Unauthorized access includes unauthorized use, disclosure, modification, destruction, and issuing of commands. It requires that access to the protected resources be controlled.

SECURITY MECHANISMS

Security mechanisms (Acharya and Muthukrishnan 1998) are means to achieve the security services described above. There is no single mechanism that can provide all the security services. How- ever, there is one main class of techniques that underlies most of the security mechanisms in use, namely cryptographic mechanisms.

Asymmetric Cryptography

The concept of asymmetric cryptography, or public key cryptography, was first introduced in 1976 by Diffe and Hellman. An asymmetric cryptosystem is a cryptographic scheme in which two distinct keys, known as the public key and the private key are used. The public key, as its name suggests, can be made public to everyone in the communications system. On the other hand, the private key must be kept secret, and known only to its legitimate owner. Although confidentiality is not important for the public key since it must be made accessible to anyone, it is important to ensure its integrity. To be precise, it must not be possible to alter a person's public key. The concept of a Public Key Infrastructure (PKI) has therefore been introduced as a means to generate, distribute and manage 'public key certificates' which are used to bind the identifier of a party to that party's public key. A widely adopted standard for the format of digital certificates is X.509. There are a number of broad classes of asymmetric cryptographic scheme, including encryption schemes, digital signature schemes, and key agreement schemes. However, only the first two will be described here since they are most relevant to this chapter.

Asymmetric Encryption

Asymmetric encryption schemes use public keys for encryption and private keys for decryption. The best known algorithm for public key encryption is RSA (Rivest, Shamir, and Adleman 1978). Specification for public key encryption, including the use of RSA, can be found and also in the emerging international standard. Asymmetric encryption can be used to provide data confidentiality.

Digital Signature

Digital signature (Agosta and Russell. 1996) as data appended to, or a cryptographic transformation of, a data unit, that allows a recipient of the data unit to prove the source and integrity of the data unit and protect against forgery. Digital signature mechanisms are asymmetric cryptographic techniques which can be used to provide entity authentication, data origin authentication, data integrity and non-repudiation services. A signature scheme consists of two components, namely a signing algorithm and a verification algorithm. The signing algorithm involves the transformation of the message into a signature, using the signing entity's private key. It should be clear that, for a digital signature scheme to work there is a need for a verification process, so that it is possible to verify whether a signature on a message was genuinely created by the claimed entity. This verification process takes as input the signature, the message, and the signer's public verification key, and outputs an indication as to whether or not the signature on the message is valid. Many digital signature schemes have been proposed over the last 25 years (Akyildiz and Ho 1956). Digital signature schemes can be used to provide data origin authentication, data integrity, and non-repudiation.

Symmetric Cryptography

Symmetric cryptography is a cryptographic scheme in which either the same key (secret key) or two keys that can be easily computed from each other are used. That is, in a symmetric cryptographic scheme, the communicating parties are required to share a key which must be kept secret. There are a number of symmetric cryptographic schemes, including encryption schemes, message authentication codes, and cryptographic hash functions (Ananda and Tay 1991).

Symmetric Encryption

Unlike asymmetric encryption, symmetric encryption uses a single key for both the encryption and decryption transformation. The encryption is said to be symmetric if, for each associated encryption/decryption key pair, it is computationally 'easy to determine a decryption key knowing only the encryption key and vice versa.

There are two commonly used types of symmetric encryption scheme, namely stream ciphers and block ciphers. A block cipher is an encryption scheme which breaks up the plaintext messages to be transmitted into strings (called blocks) of a fixed length and encrypts one block at a time. On the other hand, a stream cipher is 'an encryption mechanism such that using a running key or a fresh one-time-pad key stream, an encryption encrypts a plaintext in bit-wise or block-wise manner. As for asymmetric encryption, symmetric encryption schemes can be used to provide data confidentiality.

Message Authentication Codes

The second type of symmetric cryptographic technique we describe is the Message Authentication Code (MAC). This mechanism can be used to provide data origin authentication and data integrity services. The originator of data inputs the data to be protected into a MAC function, together with a secret key the resulting output (a short fixed-length bit string) is known as the MAC. This MAC (Arnold and Gosling 1996) can then be sent or stored with the data being protected. The verifier of the MAC simply uses the same secret key to re-compute a MAC value on the data, and the data is accepted as valid if and only if the recomputed MAC agrees with the value sent or stored with the data.

Cryptographic Hash Function

Hash functions take a message as input and produce an output referred to as a hash-code, hash-result, hash-value, message digest, or just hash. More formally, 'a hash function is a function which maps strings of bits to fixed-length strings of bits. It must also satisfy the following three properties.

it must be computationally infeasible to find for a given output, an input which maps to this output, and it must be computationally infeasible to find for a given input, a second input which maps to the same output. It must be computationally infeasible to find two different inputs which map to the same output. Hash-functions form a vitally important part of almost all commonly used digital signature schemes. There are a number of types of hash functions, for example those based on block ciphers, those based on modular arithmetic, and dedicated hash functions for cryptographic hash function standard (Awduche, Gaylord, and Ganz 1996).

SECURITY ISSUES

Electronic commerce is growing in significance. Many products, tangible and intangible, are sold over the Internet, with payments typically made by debit or credit cards. In parallel with this, there is an increase in concerns associated with the security of the payment systems used to process online transactions. Probably the main concern of most Internet users relates to the confidentiality of payment card information, since disclosure of this information to a hostile third party could enable that party to make fraudulent transactions at the user's expense. However, security for online transactions is not limited to data confidentiality, but also includes other security services such as authentication, identification, non-repudiation and data integrity.

In a typical debit/credit card payment system there are four parties involved, namely a client, a merchant, an acquiring bank (Awerbuch and Peleg 1995) and a card issuing bank. A client, i.e. the cardholder, makes a payment using a card issued by the card issuing bank (issuer) for something purchased from a merchant. The acquiring bank (acquirer) is the financial institution with which a merchant has a contractual arrangement for receiving (acquiring) card payments.

In this chapter, we consider the security requirements for each of the four parties involved in an electronic commerce transaction. Some currently proposed electronic commerce protocols are then described and briefly analyzed in terms of how well they satisfy these security requirements. Note that the analysis is based on a typical Internet payment transaction, i.e. where a user makes a payment by entering card details through a web interface, the merchant server processes the transaction using a back-end authorization system, and this enables the transaction to be sent to the acquirer and subsequently to the financial network. Further discussion on security issues for e-commerce, and details of many of the schemes described here, can be found in recent books on e-commerce, e.g.

SECURITY REQUIREMENTS

A typical card payment system involves four parties (Bach 1990), namely a card issuer, an acquirer, a merchant and a client. The security requirements for each party vary and hence they will be examined individually. However, the requirements for acquirers and issuers are combined since they are both financial institutions, they are both contractually obliged to abide by the rules of the relevant payment system, and it can reasonably be assumed that they have a similar risk model.

Issuers and Acquirers

Non-Repudiation

Issuers and acquirers need to ensure that neither clients nor merchants can deny their participation in a transaction (where the transaction may involve a refund from merchant to client). In order to achieve non-repudiation, identity authentication may also be needed.

Authentication

Client authentication is required for the issuers and acquirers so that they can prove that it is the client who authorized the payment and that he/she is a legitimate cardholder. Otherwise, a client can deny making a transaction and the issuer may end up being liable for refunding the amount to the client. On the other hand, if an electronic transaction is found to be fraudulent, merchants are liable for 'card not present' chargeback's. Therefore, it is important for the acquirer to ensure merchant non-repudiation to prevent them challenging their liability.

Integrity

It is also important to ensure that once details of a transaction have been confirmed, no one can maliciously modify them. Merchants must not be able to alter the amount that a client has agreed to pay. To be more specific, it should not be possible for a merchant to change the amount after it has been authorized by the card issuer. Similarly, a client must not be able to change the amount that has been authorized.

Replay Protection

A malicious merchant should not be able to use a once authorized transaction to obtain a repeat payment. Additionally, merchants should not be able to use an old transaction to request a new payment authorization no matter how many similar transactions the client has made with them. Issuers and acquirers need a mechanism to detect if a transaction has been replayed so that they do not authorize an illegitimate transaction.

Merchants

Non-Repudiation

A merchant needs evidence that a customer has agreed to pay the amount associated with a transaction. A merchant also needs to verify that the client is the legitimate cardholder; otherwise, the merchant can be liable for chargeback's. This occurs when a client tells his/her issuer that a particular transaction was not made. The card issuer then immediately submits a chargeback to the acquirer to recover the amount from the account of the merchant in question. Within a predefined period of time, the merchant can dispute the chargeback by providing evidence of, for example, purchase or delivery. Therefore, it is important for merchants to have non-repudiable evidence of the transaction, i.e. to have client non-repudiation. Furthermore, an issuer should not be able to deny having authorized a payment.

Authentication

As stated before, merchants need client authentication to make sure that the client is the legitimate cardholder. Moreover, they need to be sure that they are communicating with the genuine acquirer. Otherwise, an adversary may masquerade as an acquirer and authorize an illegitimate transaction.

Integrity

No one should be able to change the details of a transaction once they have been agreed upon. A merchant will not wish to be credited with payment for less than the amount agreed. In addition, an acquirer or issuer should not be able to modify a transaction that has been authorized.

Replay Protection

A malicious client should not be able to present an old proof of purchase to claim for repeat delivery of goods. Likewise, it should not be possible for an acquirer to claim that a merchant has obtained a payment using an old transaction.

Clients

Confidentiality and Privacy

Transaction confidentiality, especially card information confidentiality, may be the security service of most concern to users. It is important that cardholder account details are kept secret from any party except the issuer, since they are the main basis on which Internet payments are made. Moreover, some users may require confidentiality protection for the nature of their transactions.

Integrity

As for the other parties, transaction integrity is important to the client. No one should be able to maliciously modify the transaction details once they have been confirmed. Clients will not want an adversary to change a delivery address, the price, or the description of the merchandise after they have agreed a payment.

Authentication

A client needs to be sure that he/she is dealing with a trustworthy merchant. When shopping on the Internet, it is relatively easy to be lured into visiting a site which appears to sell something but is actually simply collecting card details. Even though a client may have made a purchase from a site before, it is not always obvious whether the page that is being fetched is authentic.

Replay Protection

Clients need a mechanism to ensure that a malicious merchant or an adversary will not be able to reuse previously authorized payments to make a repeat charge.

Non-Repudiation

Clients also require non-repudiation, for example a proof of payment so that no one involved in the transaction can repudiate that a payment has occurred.

MEETING THE SECURITY REQUIREMENTS

In this section, we examine various currently available security protocols. A short description of how each protocol works is given, followed by a review of the security services provided.

SSL and TLS

The Secure Sockets Layer (SSL) protocol was launched in 1994 by Netscape, with the primary goal of providing secure communications between web browsers and web servers. Security services provided include server authentication, data confidentiality, (optional) client authentication and data integrity. In 1995, the Internet Engineering Task Force (IETF) introduced a similar protocol named Transport Layer Security (TLS) version 1.0. SSL and TLS are by far the most widely used protocols providing security for transactions made over the Internet. Because of their importance, the whole of Chapter 4 is devoted to a detailed analysis of both SSL and TLS.

Secure Electronic Transaction (SET)

The Secure Electronic Transaction (SET) protocol was developed by Visa, MasterCard and various computer companies to facilitate secure electronic commerce transactions. The protocol provides confidentiality of payment card details, data integrity, authentication of both merchant and cardholder, and the ability to validate or authorize transactions. SET extensively employs public key cryptographic techniques to provide such security services. As a result, one of the most important prerequisites for the protocol is that all the parties involved must have their own distinct key pairs with corresponding public key certificates. Moreover, cardholders have to install special software before they can start using SET.

SET is a complex protocol involving more than ten steps for each transaction, and it is not the aim of this chapter to provide full details of the protocol's operation. In brief, the protocol requires every participating party to cryptographically sign transmitted messages. Sensitive information is also encrypted using a secret session key. One of the most innovative features of SET may be the use of a 'dual signature' which allows merchants to verify the integrity of the order information yet not see the card details. SET appears to be a well designed protocol that aims to provide a high level of security for Internet transactions, and that satisfies all the security requirements outlined previously. Unfortunately, SET has not been adapted to any significant extent indeed, it is not clear whether it will ever become widely used. One of the most important obstacles to SET implementation is that the protocol is so complex that it is difficult and costly for the parties involved to implement it. Moreover, the use of public key cryptographic techniques is costly in terms of computational overhead, performance, and the Public Key Infrastructure needed to support it. As a result, the benefits of security that SET gives may not be sufficient to bring about its adoption.

Visa 3-Domain Secure

The 3-D secure protocol has recently been developed by Visa. The protocol aims to provide cardholder authentication for merchants using two types of servers: Access Control Servers (ACSs) operated by card issuers and the Visa Directory Server. The cardholder must enroll with his/her issuer ACS before using the service.

When a transaction is to be made, the merchant server queries the Visa Directory Server, using a Merchant Plug-In, to determine whether the cardholder has registered for the authentication service. If so, the Visa Directory Server returns the web address of the appropriate ACS to the merchant. The merchant server then redirects the cardholder browser to that ACS which will in turn prompt the cardholder to authenticate him/herself using a password and/or a Visa smart card. If the authentication process is successful, the ACS redirects the cardholder browser back to the merchant server, which can then proceed with the traditional authorization process. The Visa 3-D Secure also employs another server called the Authentication History server to log all the authentication attempts of each cardholder.

It is clear that the protocol does not meet all the security requirements. Indeed, it does not attempt to do so, since its aim is only to provide cardholder authentication and the associated non-repudiation to reduce or eliminate the risk of card-not-present charge-backs, i.e. where the merchant is required to take liability for a disputed transaction if the card was not physically present at the merchant premises at the time of the transaction. The protocol is also kept simple to facilitate implementation.

MasterCard Secure Payment Application

The MasterCard Secure Payment Application (SPA) is a security solution for securing payments between Merchants and Issuers for card-not-present transactions via the Internet i.e. where the card does not present at the merchant site when the transaction is made. As for Visa 3-D Secure, the scheme is designed to provide cardholder authentication and hence reduce chargeback's.

CONCLUSION

In this chapter, gives the brief details about Security and trust were identified as essential elements that need to be considered in e-commerce programs. Electronic banking is an appropriate environment for the research because banking activities contain exchange of sensitive personal records to authorized users. This makes believes of security and customer satisfaction for e-banking. Customers interact with different safety mechanisms at every stage during their online fund transfer with the banking system. The protection of data changed into the factor of security that increases positive impact on trust.

This implied that users issues related to privacy of records are the very important and most assurance on all issues related to privacy. Online banking interfaces need to talk with the clients with their techniques used by them to ensure the privacy of the consumer's information and transactions. Thus, maximizing attempt to improve the perception of privacy would help increase the extent of believe in e-banking programs. In this chapter, therefore, defined the security offerings of relevance to this online banking and after that the security mechanisms which can be used to offer the security offerings are then tested. The next chapter gives the brief details about different Attacks and Authentication for Mobile Banking.

REFERENCES

Accetta, Baron, Bolosky, Golub, Rashid, Tevanian, & Young. (1986). Mach: A new kernel foundation for UNIX development. *Proceedings of the USENIX Summer Technical Conference*, 93-112.

Acharya & Muthukrishnan. (1998). Scheduling on-demand broadcasts: New metrics and algorithms. *Proceedings of the 4th ACM/IEEE International Conference on Mobile Computing and Networking*, 43-58.

Acharya, Franklin, & Zdonik. (1995). Dissemination-based data delivery using broadcast disks. *IEEE Personal Communications, 2*(6), 50-60.

Agosta & Russell. (1996). *CDPD: Cellular Digital Packet Data Standards and Technology*. McGraw Hill.

Akyildiz & Ho. (1996). On location management for personal communications networks. *IEEE Communications Magazine, 34*(9), 138–145.

Ananda & Tay. (1991). ASTRA - An asynchronous remote procedure call facility. *Proceedings of the 11th International Conference on Distributed Computing Systems, 26*(2), 172–179.

Arnold & Gosling. (1996). *The Java Programming Language*. Addison Wesley.

Awduche, Gaylord, & Ganz. (1996). On resource discovery in distributed systems with mobile hosts. *Proceedings of the 2nd ACM/IEEE International Conference on Mobile Computing and Networking*, 50-55.

Awerbuch & Peleg. (1995). Online tracking of mobile users. *Journal of the ACM, 42*(5), 1021–1058.

Bach. (1990). *The Design of the UNIX Operating System*. Prentice-Hall.

Tan & Sutherland. (2004). Online Consumer Trust: A Multi-Dimensional Model. *Journal of Electronic Commerce in Organizations, 2*(3), 40-58.

Yousafzai, Pallister, & Foxall. (2005). Strategies for Building and Communicating Trust in Electronic Banking: A Field Experiment. *Psychology & Marketing, 22*(2), 181-201.

Chapter 2
Attacks and Authentication for Mobile Banking

ABSTRACT

It is important that any new product or service is based on user needs and requirements rather than being driven by technology. This is an important issue to consider with SMS and mobile banking. Research related to the usability design issues surrounding online banking are relevant to SMS banking. Concepts related to authentication, online banking processes, and other relevant information security issues are discussed further. Literature relevant to these factors is also considered throughout this process of reflection. This chapter discussed the concepts related to authentication, online banking processes and other relevant information security issues are discussed further. Literature relevant to these factors is also considered throughout this process of reflection. In next chapter discusses about Web based electronic money for online banking.

INTRODUCTION

Anywhere, any time, on any device, that turned into throughout of current era of computer science even though any device additionally to mobile devices. Mobile devices lacked the processing energy and had gradual information connections to carry out easy operations. Today we see a resurging mobile revolution. The revolution on mobile gadgets that are very effective processors and fast data communication. One of those promises changed into banking

DOI: 10.4018/978-1-5225-2759-6.ch002

the usage of a mobile tool. It delivers the client the convenience of banking anywhere and every time, it additionally offers the financial company and other services. Offering net banking on a cell device introduces its own specific demanding situations that have to be handled, as an example security issues, which can undermine to accept the true information with clients have in their bank account. Trust is one of the fundamental cornerstones of banking. Without it, it might be difficult for banks to access in commercial enterprise. However, the customer also expects traditional banking offerings on mobile devices.

MOBILE BANKING

The numbers of mobile devices in use retain to upward thrust for instance, with around 6 billion mobile telephones currently in use. It is difficult to complain with the achievement of mobile devices. Mobile phones have turn out to be herbal extension of the individuals who use them in a extensive variety of activities. Not simplest did mobile telephones change the way people speak with each different situations, however they also changed the way facts is processed and the way business is conducted. Before the security issues of mobile banking can be mentioned, a high level in order to accessing banking providers on a cellular device gives customer freedom. Not most effective does it placed them in price, it also allows them to do their banking independently of their vicinity and time. In conventional banking, a bankers wish to be provides at one of the branches of a financial institution and has don't forget the opening hours the financial institution has set up. Mobile banking gives the client every other degree of freedom, particularly geographical independence. We see that banks have detained the vast use of mobile devices as an opportunity to connect with their clients. Several sort of produce may be presented, as an example a manner to control financial plan the use of a mobile tool, or as a digital wallet. A uncontaminated difference may be made between mobile price and mobile banking, It define mobile fee and banking as 'the use of mobile telephones to pay for services (bus, train, movies, entertainments), items (retails stores, coffee stores, eating places, vending machines, online stores), payments (electric, fuel, credit score playing cards, phone), and switch price range (bank to mobile, bank to financial institution, cellular to mobile)'. It is essential to notice the difference between mobile price and mobile banking. In the context of shifting budget, mobile banking is a subset of e-banking and mobile banking can be described as that kind of execution of economic assistance in the course of which, the purchaser makes use of mobile communication strategies in conjunction with mobile devices.

Mobile banking has several advantages for the customer. It allows customers to carry out their banking supports to anywhere, any time at a lower price. Furthermore it offers the customer acquire to access the information applicable to them, be it a request for the modern financial credit stability or an overview of recent transactions. Traditionally one might have to call or visit the bank to access the current account balance, or the status of a transaction.

AUTHENTICATION

Authentication (D. R. Cheriton and W. Zwaenepoel 1984), in general, can be divided into two categories: Entity Authentication and Data Authentication. Although, we often consider these two categories to be quite similar, they are very different when it comes to why and how these authentications are carried out. While entity authentication always begins the process, data authentication comes into play after the user authentication is completed. With this said, it is important to understand that although entity authentication and data authentication are two major modes of authentication, there are different methods used to employ such authentication tactics.

Entity Authentication

Entity authentication is the process of identification of correct users by the service providers and vice-versa. Therefore, an identity of a person or an organization must remain a unique factor, which is used to distinguish between different persons or organizations in particular domains. Entity authentication typically takes place at the beginning of a session or interaction. Entity authentication can be divided into three modalities: syntactic, semantic and cognitive authentication described further below.

Syntactic Authentication

The verification process between two entities where a source entity transmits its authentication credentials such as passwords to a recipient entity. The recipient can then validate the authentication credentials. The recipient is indifferent to the authenticated entity, and does not apply any policy.

Semantic Authentication

The verification process between two entities where a source entity transmits its authentication credentials to a recipient entity that validates it. However, in semantic authentication, an additional step of compliance testing with a security policy is also done.

Cognitive Authentication

The verification process between a cognitive source entity and a recipient entity by transmission of authentication credentials which is validated along with a compliance testing to a security policy. In addition, this is done through, a user friendly representation of identity attributes which enables the cognitive relying party to recognize and reason about policy compliance of source entity.

There are various means of authenticating human users, such as username/ password combinations, and biometric testing. However, user authentication alone is insufficient to secure online transactions; therefore, data authentication is also required. The concept of identity management is a central issue in the current context of rapidly increasing online services. There is a huge increase in the number of online services provided through the Internet.

With this increase of services, there are also a greater number of users who use them. Identity management under such a context becomes a challenge for the service provider, as well as for the user. The service provider must issue credentials and identifiers to the users, and when service operations are initialized by a request from the user, the service provider must validate user credentials and provide allocated functionalities and services for users. Most of the identity management systems are service provider centric. In this sense, a need for a user-centric identity management system was discovered.

The first thing necessary to understand in a user-centric identity management system is the use of identifiers and credentials by a user. A user depends on a more manual means of managing all the credentials for different service providers (David Clark 1985). Memorization of all these credentials and identifiers for different service providers is a daunting task indeed. Therefore, a new approach towards management of identity from the user's perspective is required. This particular concept gave rise to a concept of Personal Authentication device (PAD). The PAD can validate a user with a PIN, after this process the PAD can be activated and used by a user for personal purpose, which makes it user-centric. The PAD can have cross-platform authentication abilities that can accommodate different authentication protocols. It could be connected via communication channels such as Bluetooth or Wireless Local Area Network (WLAN). However, a dedicated device such as PAD can be difficult to carry in comparison to a mobile phone, which is already carried by almost everyone. Mobile phones have greater capacities in terms of computing, processing, transmission and security. However, with the tough competition in the mobile market, where usability and ease of access of mobile devices outweigh the need for security features, it becomes difficult for manufacturers to make normal mobile phones highly secured. The French company Taz-Tag has introduced a mobile phone (TPH-ONE) based on Android which also integrates a secure element which can be accessed in a secure state. Unlike other mobile phones including Taz-Tag, The Off PAD does not provide similar services as mobile phones such as web browsing, music or making phone calls, therefore, it will not serve any another purpose than authentication. In this sense, the Off PAD is targeted towards commercial adoption by those who are in need of high-security measures, however, some limitations in the form of economic viability are of some concern. The Off PAD can be comparatively more expensive than OTP (One Time Password), but the overall security provided by OffPAD and its services can outweigh this disadvantage.

One solution to the problem of memorizing passwords for different browsers and browser enabled websites are the storage of such passwords within the browser's security umbrella. This creates a dependency on the security provided by the service provider for authentication, whereas the Off PAD enforces the user to remain in control of all his/her credentials.

While user authentication is just one part of the solution, another part lies in the process of service-provider authentication. With this said, it is necessary to understand that unlike user authentication, service provider authentication consists of several complexities by nature. These complexities come in the forms of the scope of service providers, which can be global or local, stability and reliability. Identity theft is a common problem in today's world, where most of the transactions are done online. The identity of the user can be manipulated either by malware in the client platform, using a Man-in-the-Middle within the Internet or at the client. Therefore, to verify the identity of the user, the authentication process must start from the user-side itself regardless of the presence of any malware. In the client, this technology is based on a device called Off PAD (Offline Personal Authentication Device) used as authentication device to perform trusted transactions.

The Authentication process has to gradually evolve along with the evolution of the entire computer system. With this fact, authentication is not just concerned with single or multiple human users, but is also related to automated systems. Therefore, there are two different types of entities, first is the human entity where there are users or organizations, and then there is another group called system entities which consist of clients or server systems.

Data Authentication

Although there is already considerable amount of research on user authentication, however, data authentication cannot be neglected. There are two types of authentication services according to International Telecommunication Union (ITU-T):

Data Origin Authentication

Data origin authentication is described as the corroboration that the source of data received is as claimed (ITU-T, 1991). Data origin authentication does not provide any protection against duplication or modification of data. Its authentication remains limited to assurance of the source of the data is legitimate.

Peer Entity Authentication

Peer entity authentication is described as the corroboration that the peer entity in an association is the one claimed (ITU-T, 1991). Its services provide assurance regarding the legitimacy of the involved entities in network sessions. Its use is specific to the confirmation of identities of involved entities from initialization of transaction or during the data transmission phases. It is specific to avoiding unauthorized access.

X.800 Recommendation: ITU-T

X.800 is recommendation security architecture from ITU-T for OSI (Open Systems Interconnection). It is one of the sought after standard by international security vendors. It focuses on three factors:

Services

Security service (W. Paul Cockshot. Ps-Algol Implementations 1990) ensures the security of system and data. It is a service dependent on a protocol layer of the open communication system. X.800 services consist of five categories:

Authentication

Authentication is the service provided for authentication of a peer-entity or a data source. Peer Entity authentication services are provided by the (N)-layer, which ensures to the (N+1) entity that peer-entity is the claimed (N+1) entity (ITU-T, 1991). ii. Data origin authentication is provided by the (N)-layer, which ensures to the (N+1) entity that source of data is the claimed (N+1) entity (ITU-T, 1991).

Access Control

In OSI, this particular service is targeted towards unauthorized access of resources. These resources may or may not be OSI dependent resources. This service is implemented at the early stage where secure entities are identified, and unidentified entities are denied access.

Data Confidentiality

It consists of four types of confidentiality parameters, which protect data from unauthorized access and disclosure.

1. **Connection Confidentiality:** It provides security to data specific to users during specific connections.
2. **Connectionless Confidentiality:** It provides security to data specific to users during single connectionless service data units.
3. **Selective Field Confidentiality:** It provides security to selective portions of users data specific to connections or on single connectionless service data units.
4. **Traffic Flow Confidentiality:** It provides security to data, which might be extracted from the flow of traffic.

Data Integrity

It is the security (E. C. Cooper 1987) of data from origin to transmission, which means it provides assurance regarding the source of the data as well as the accuracy of data throughout the transmission process. It consists of five different forms of security aspects:

1. **Connection Integrity With Recovery:** It provides integrity services for all user data on specific connections by detecting any sorts of modifications such as insertions, deletions, updates or reiteration of any data.
2. **Connection Integrity Without Recovery:** It is similar to connection integrity with recovery besides the fact that it does not provide recovery services.
3. **Connectionless Integrity:** It provides integrity services for a single connectionless single data unit which checks for any modifications and reiterations along the transmission process.
4. **Selective Field Connection Integrity:** It provides selective integrity on specific fields of the user data over specific connection and checks for any modifications such as inserts, deletes or reiterations.
5. **Selective Field Connectionless Integrity:** It provides selective integrity on specific fields of the user data over a single connectionless single data unit and checks for any sorts of modifications.

Non-Repudiation

It consists of any of two forms proof of origin or proof of delivery, which is described below:

1. **Non-Repudiation with Proof of Origin:** The proof of origin of data is provided to the recipient, which protects against any false claims of denial of sending data.
2. **Non-Repudiation with Proof of Delivery:** The proof of delivery of data is provided to the sender, which protects against any false claims of denial of receiving data.

Mechanisms

Encipherment

It provides security through confidentiality of data or traffic. It works on the principle of use of ciphers which are basically codes used to encrypt and decrypt messages. These ciphers are of two types:

Symmetric (Shared-Key Cryptography)

In symmetric cipher, the same keys are used to encrypt and decrypt by sender and receiver. The user is in possession of a secret key, which remains unique as well. The user can then authenticate him by sharing this secret key to the authentication server. The user sends the username and a randomly generated number or code, which is encrypted using a secret key to the server. The server consists of the shared secret key, which is used to match the received encrypted code, and if the match is absolute, the user is authenticated. OTP (One Time Password) tokens are a tangible implementation of symmetric-key.

Asymmetric

Asymmetric cipher, which is also known as public-private key cryptography, uses two sets of keys. Unlike symmetric cipher where the same keys are used, here the public keys are known but the private key remains secret to the specific user. This creates a higher level of security.

Digital Signatures

It consists of two procedures:

1. **Signing of Data Unit:** It is done using two methods:
   a. Encryption of data,
   b. Creation of a cryptographic key or check value for the data.
2. **Verification of Data Unit:** It is done using a publicly known process to validate the signature, which remain private to its owner.

Data Integrity Mechanisms

The integrity (Helen Custer 1993) of data is related to either single data unit or a stream of data. The integrity of single data is maintained by placing checkpoints at sending entity and receiving entity. The sending checkpoint assigns a check-value or key to the data while the receiving entity produces exact check-value, which ensures the integrity of data. However, for connection-mode data transmission, additional forms of integrity checks are required such as time-stamping, sequence numbering or secondary encryption. This creates a secondary layer of integrity check to protect against several threats such as replay, loss, disorder or mismatch.

Authentication Exchange

This authentication mechanism consists of three different techniques:

1.   Use of authentication information such as passwords
2.   Cryptographic techniques
3.   Use of characteristics or possessions of entity

Traffic-Padding Mechanism

This mechanism is used to provide protection against traffic pattern analysis. Using confidentiality mechanisms can protect it further.

Routing Control Mechanism

This mechanism works by pre-arranging a secure route for data transmission, detection of unwanted data forbidden under security policy and re-routing such data or deny access of data to specific networks.

Notarization Mechanism

This mechanism utilizes a third-party notary service to validate the authenticity of the transaction. All involved parties trust the notary service. Integrity mechanisms such as digital signatures, decipherment etc. as directed by the notary is used.

Attack

According to X.800, in general, there are two different types of attacks:

Active Attacks

These are attacks targeted at changing the state or operation of the system. Some active threats are:

1. **Masquerade:** A false entity presents itself as the claimed entity. It usually consists of other malicious attacks such as replay or modification of data.
2. **Re-Play:** Re-play occurs when the same message is repeatedly used by different entities for unauthorized access
3. **Modification of Messages:** Modification of message is an unauthorized alteration of a transmitted message, which gets changed on the way.
4. **Denial of Service:** The process of suppression or generation of messages transmitted from source to destination in such a manner that it disrupts the services on which the system depends is called denial of service.

Passive Attacks

X.800 defines passive attacks as a threat of unauthorized access to information without changing the state of the system (ITU-T, 1991).

If these threats are discovered on time, then they cannot cause problems.

1. **Wire-Tapping or Eavesdropping:** Wiretaps are used to perform passive observation of data on different channels used for transmission, which can cause a leak of information sensitive in nature such as financial and personal data.
2. **Traffic Analysis:** The observation of flow of traffic to understand patterns and inference of data from such patterns can be damaging. Traffic analysis can cause threats to the cryptographic system as well which depends on specific trends or patterns.

Threats Specific to Online Banking/E-Commerce

Online banking (Sajal K. Das, Sanjoy K. Sen, and Rajeev Jayaram 1996) is a feature provided by almost every bank to provide its customers with essential services. With the advent of the Internet, like every other industry, banks also quickly to accept its advantages. Online banking quickly became an integral part of the banking industry as well. The reason behind such a strong response to online banking is because of the customers. Customers of the banking industry became the true owners of their bank accounts. However, with the increasing use of cell phones and the expansion of 3G and 4G wireless services, a new front on the battle for customer satisfaction is opened. Online banking is now moving towards mobile online banking.

These changes have brought significant effects in our personal and professional lives as well. But, like every coin has two sides, there are several problems as well. One of the biggest issues when dealing with online banking is security. In one study done in Singapore, Internet banking users accepted it because it seemed convenient, simple and more compatible. Banks are a large source of personal information. However, in a 2005 study in the United Kingdom (U.K.), the banks were found to provide insufficient security mechanisms for its customers. Banks have however tried several solutions to solve problems related to information security in different ways. Banks such as HSBC gave away antivirus software to its clients for free. Banks have also come up with some high tech solutions of confirmation of information by using a cellular network such as SMS or through email confirmations. They have used two-factor authentication techniques such as PIN code generators. Now, the trend is growing towards the use of multi-factor authentication techniques such as biometric devices, pattern based encryptions, image capture etc.

Phishing: Identity Theft Attacks

This particular problem of phishing is highlighted with the fact that this is one of the most employed techniques by fraudsters to obtain credentials from online banking customers to steal financial information (Vittoria de Nitto Persone, Vincenzo Grassi, and Antonio Morlupi 1998). Some infamous malwares which are used for phishing attacks are ZeuS and Spyeye. Phishing works through the concept of deception and infection. They portray themselves as genuine messages or warnings, which deceive the user to believe that they are in genuine danger of some sort. Then they ask the user to download some executable file hidden under genuine software to infect the device and provide them with their credentials.

Man in the Middle: Channel Breaking Attacks

Online banking under present conditions, utilize two-factor authentication as a solution to a number of security problems. Two-factor authentication depends on the concept of improbability to estimate the randomization process, where a random password is generated to a specific user. However, the user remains the static element in this transaction. The two-factor authentication works very well for passive attacks such as wiretapping, dictionary attacks, password guessing.

However, the change of attacks from passive to active attacks adds a different degree of vulnerability to use of two-factor authentication techniques such as OTP (One Time Password) or PGT (Password Generating Token). Man in the middle attack is an example of such an attack, where a fake website which has almost similar web address as the original address can be used to fool potential customers to provide their credentials. These credentials can be used to access the original website. The attacker can misuse these credentials in a variety of ways. The attacker depends on the user to complete the authentication process, including two-factor authentication such as SMS verification or OTP verification. After the completion of two-factor authentication, the attacker can start the attack. One way to mitigate issues of man in the middle attack is through the use of session ids. However, attackers have discovered a new way of attack called "Man in the Browser" discussed further.

Man in the Browser: Content Manipulation

MitB (Man in the Browser) attacks are possible through several techniques such as Browser Rootkit. This is a popular technique employed for the Man in the Browser attacks. This attack usually occurs between a user and the browser. The browser is compromised using various malicious browser extensions, which can alter functionalities within browsers. These extensions attach themselves to various parts within the browser including the installation files located locally within the hard drives. Man in the Browser attacks are more sophisticated in the sense that they give the user the impression of security, as everything seems normal for the users. MitB attacks employ browser extensions as these extensions have different privileged capabilities to enhance or alter services provided through operating systems. These attacks utilize browser helper objects, JavaScript, AJAX, Browser API and DOM (Document Object Models). The attacker can remain in control of different sessions and these connections can run on TLS/SSL connections as well.

SERVER AUTHENTICATION

These above-mentioned techniques could apply for users as well as servers. Some scenarios for such an attack can be:

- **Error While Typing the Domain Names of the Financial Institution:** An error while typing the domain names of the financial institution can be used as an opportunity by these phishing websites. Such an error can redirect the user to a cloned website.
- **Security of Servers or Websites:** A lack of security for servers or websites can severely be misused by the phishing messages as well as hosting purposes of illegitimate websites.
- **Offline Dictionary Attacks:** A list of passwords can be used to attack in offline mode.
- **Online Dictionary Attacks:** An Online dictionary attack is the same as an offline dictionary attack, however, it varies in the context of verification of passwords in online mode to find out if it is right or not.
- **Online Identity Impersonation:** An Online identity impersonation is one of the issues with current social media. Facebook, for example can be used to create a fake account by establishing a fake user. There is no particular possibility for Face book to link the user to the corresponding user profile which means it cannot verify whether the user is as claimed. Anyone can upload a photo of someone else on Facebook and there are no means to verify if that photo is linked to the real user or not. Public Key Infrastructures and Pretty Good Privacy are in some sense solutions to provide a link between online entities to an individual. However, these solutions lag behind, in terms of issuing personal certificates to every individual. Certification Authorities cannot provide individual certificates to all users.

- **Compromise of Certification Authorities:** Certificates are a means of assurance that the information is true. Certificates are mostly distributed by various third party elements that are known as Certification Authorities (CAs). The validity of such certificates rely on a cryptographic watermark however, valid certificates can be used for phishing attacks. So the validity of certification does not always constitute in the valid website itself. So this creates vulnerability for CAs and their certificates as well.

- **Server Certification:** Widely done using WTCA (Web Trust for CA), this particular dependency upon CA creates a security assurance limited to the capabilities of the CA. Therefore; the server will only be as secure as its strongest CA.

- **Information Variation at Servers:** In order to manage information within the security procedure of an organization, the information is classified according to its sensitivity, confidentiality, and integrity. This information is then stored in a relevant server, which consists of the required security procedures for access. This is done for the availability of information. Server authentication is becoming more troublesome as the Internet gets broader and the number of users increases rapidly. The most implemented process of authentication of online entities during online transactions is TLS (Transport Layer Security) previously known as SSL (Secure Socket Layer). TLS is responsible for the security of transaction and encryption of transaction related information. However, the password used for user authentication can be a point of vulnerability created by the use of phishing websites, which is one of the biggest threat for online banking users. Some of the other threats mentioned above, such as dictionary attacks, are also possible in such scenarios.

In early days, passwords, in general used to be simple and short but however, with present security needs, users are recommended by service providers to create long and complex passwords with alphanumeric characters, which makes them difficult to remember unless someone has an exceptional memory. Public Key Infrastructure (PKI) is another solution to such as problem; however using such an elaborate method of security has its set of limitations in terms of offline attacks on the system, where the system is limited to the user credential storage. User is limited to using a system which has the keys installed. Such a limitation makes it easier for attackers to focus their attacks on a limited number of systems.

A few years back, online banking relied heavily on the username password combination, however, with new threats, the use of one-time password along with username, password combination has increased. Therefore, the trend is moving towards multi-factor authentication, where biometric authentication techniques are employed. Almost 52 percent of financial organizations in Asia use fingerprint security systems.

TYPES OF SERVER AUTHENTICATION

There are two popular cryptographic transport protocols (S. Deering and R. Hinden 1998) called TLS and SSH which use RSA or DSA as primary authentication techniques. For server authentication, two of the most important cryptographic transport protocols are:

1. **TLS – Transport Layer Security:** In the case of a TLS encrypted with RSA, the server sends a TLS certificate and the public key to provide credentials for a handshake. An attack can occur at this point if the attacker is in possession of the server's private key and the message containing valuable information including session-controlled information can be decrypted.
2. **SSH – Secure Shell:** SSH consists of two different versions of handshake:
    a. **SSH – 1:** In SSH - 1, the client can encrypt the session key using the public key of the server.
    b. **SSH – 2:** In SSH - 2, a session key can be generated from Diffie-Hellman key exchange.

In SSH - 2, a session key can be generated from Diffie-Hellman key exchange.

Similar to TLS, an attack can occur when an eavesdropper with the private key of the server can decrypt the SSH - 1 messages. While SSH -2 provides a greater level of security using Diffie-Hellman, but it is still open to active attacks.

SSH enables authentication of the server to a client using signature verification while the handshake takes place. The channel is encrypted however, the password is in plain text and if the attacker already has a private key of the server, then the password is compromised.

Domain Name System (DNS)

DNS (Domain Name System) can be considered as one of the important factors for the success of the Internet in general. Let us consider us remembering all the Internet addresses by memorizing the digits, which would impossible. DNS is a convenient solution to this problem. DNS is a link between a human user and the server.

DNS is a hierarchical distributed system spread around the world. It consists of TLDs (Top Level Domains), which remains at the highest level followed by SLDs (Second Level Domains), which give names to domains. Various Internet applications, including emails and surfing the web, depend on DNS. The number of host records has grown from 20,000 in 1987 to 1,033,836,245 in July 2015 i.e. more than 1 billion.

Although DNS has significantly increased in size to exponential size, there are security glitches in DNS. These errors are concerned with integrity and authenticity. As described above, DNS is hierarchical in nature and there are a large number of domain names registered to different organizations including various people. The primary use of DNS is its ability to distinguish one website from another. ICANN (Internet Corporation for Assigned Names and Numbers) is an organization, which governs the policies related to DNS and implementations. Thus, it has established seven new generic top-level domains (gTLDs). An extensive study of DNS by Danzig et. al. was performed on the DNS traffic on ISI (Information Sciences Institute) root name server in 1992. This study suggested an array of problems related to recursion loops, poor failure detection algorithms. Another study by Brownlee et al. suggested passive forms of attacks directed to F-root server such as queries being repeated to private address space, or invalid top level domains, source port zero and requests used to update root servers. Root servers have a relatively higher level of security. However, there is an issue of authoritative servers. These servers are more vulnerable in terms of security with regards to their location, usage and performance issues.

DNS Working

DNS is a distributed set of addresses. DNS (Richard P. Draves 1990) is normally a database of these addresses, which consists of mappings from a domain to different IP addresses. Domain names are hierarchical in nature, for example, a domain name "A" and another label or domain "B" exists then

"A.B" is considered a sub-domain of 'A'. This sub-domain is considered as a part of the 'A' domain name space as well.

Mappings related to respective domains are hierarchical as well. NS (Name Server) provides mapping of each domain. Name servers of respective domain are identified by a DNS mapping of type NS from the domain name to the corresponding domain name of the name server. Mappings of any domain name are considered legitimate only if they are received from a name server of that domain or its parent domain. Clients, on the other hand, use resolvers in order to find resource records for a certain domain. These resolvers interact or query the name servers to locate resource records. The name server then responds with a lookup and finds the corresponding resource records, or if it does not find a match, then a corresponding message is retrieved. Resolvers cache the DNS responses as per the time to live (TTL) field of response, which is normally represented in t seconds.

DIFFERENT TYPES OF DNS-BASED ATTACKS

DDOS (Distributed Denial of Service) Attacks

Denial of Service attack is targeted normally towards a single victim, where the attacker can take up all the resources of the network. DDoS attacks are normally accompanied by fake IP (Internet Protocols) addresses. This makes the attack harder to trace. One particular method of this kind of attack is the use of flood attack. In an SYN flood attack, the TCP (Transmission Control Protocol) is involved in a half connection. A random source address is created for TCP segment and TCP SYN flag is set for each TCP and requested to the server. The server then receives TCP SYN segments; the server then stores all the state information for each TCP connection by creating a TCB (Transmission Control Block). After this, the server sends a TCP SYN + ACK segment to the source address and waits for the ACK segment. However, instead of sending ACK segment, the attacker sends multiple SYN segments to the server. This will eventually create a large amount of TCB and creates a denial of service.

DNS Hijacking

The accuracy of DNS is one of the preliminary factors to affect the proficient use of the web. DNS hijacking is a real threat to allow users to accurately navigate the web and this inaccuracy leads to security issues. One such example is cache spoofing, where an attacker can insert the false address in a cache of DNS within a server or even a single browser on a client system. The most important we need to understand is that because of logistical issues, we depend on multiple DNS servers to surf the Internet. If we are looking for particular website, we will receive multiple suggestions from servers. DNS spoofing which is another name used for DNS hijacking is also transferrable through this process which makes it, even more, dangerous. An attacker can come in control of the DNS server and this can severely affect security.

After DNS hijacking, legitimate users of legitimate websites might be redirected unknowingly into a fake website. Among several mechanisms used to tackle the issue of DNS hijacking, challenge-response mechanism seems to be employed more. However, this mechanism becomes ineffective against man in the middle attacks.

Countermeasures Against DNS Attacks

SYN flooding attacks as a countermeasure to SYN flooding attacks, an approach to reduction of the TCP connection to its hash is done. This approach of reduction, reduces the size, however, it does not have the intended effect since the size is not reduced to zero. The effect is thus limited. SYN cookies embed the data instead of storing the hash for any TCP SYN + ACK segment and forward it to the attacker. This technique has the intended effect of zero state.

Although the intended effect is achieved, there are several drawbacks. One of the side effects of using this solution is that the ACK segment including the legitimate connection is received and the TCB data is regenerated which can be instantiated. TCB data contains 32-bit sequence number field, so some TCP options might not fit. TCP SYN + ACKs might not be retransmitted in response to this and the server cannot detect legitimate client ACK segment.

DNSSEC (Domain Name System Security Extensions)

It is considered to be one of the most effective means to protection against threats towards DNS. It achieves the protection using digital signatures. The DNSSEC ensures the integrity of response through cryptography. The reason to use DNSSEC to sign DNS records is to assure that the records are valid to any authenticating DNS resolver.

DNSSEC provides authority of origin, data integrity and authenticated denial of existence. It is proven effective against one particular type of DNS attack called DNS spoofing. DNSSEC can secure an entire zone through zone signing. This signing of a zone with DNSSEC creates validation to a zone without affecting any mechanism for DNS response and query. When a client sends a request or a query towards a DNS server for the DNSSEC signed zone, the DNS server sends the DNSSEC records and an attempt to validate response with the records. A DNSKEY resource record is used by a recursive DNS server to validate responses from an authoritative DNS server. This validation occurs through decryption of digital signatures, which are contained in DNSSEC-related resource records. The hash values obtained from them are computed and compared. The result of the hash values must match in order for the request to be materialized into specific data as requested, otherwise, no further response is provided except for a service failure message.

Resource Record Type Description

- **RRSIG:** RRSIG (Resource Record Signature) records contain the signatures generated with DNSSEC. A query issued by a resolver for a name is returned with a response containing RRSIG **NSEC** NSEC (Next Secure) records prove non-existence of DNS names. It is targeted towards spoofing attacks which are targeted to mislead clients to believe that the DNS that they search do not exist.
- **NSEC 3:** NSEC 3(Next Secure) is an update over NSEC which adds an additional benefit of avoiding recurrent NSEC queries. This phenomenon of recurrent NSEC queries called "Zone Walking" is prevented with NSEC 3. However, either NSEC or NSEC 3 can be used but not simultaneously.
- **NSEC 3:** PARAM NSEC3PARAM (Next Secure 3 Parameter) determines the NSEC 3 records to be included in the responses to non-existent DNS names.

DNS key DNSKEY record is used for validation process by DNS server.

HOW DOES DNSSEC WORK?

Step Query-Response Optional DNSSEC Data

1. A DNS client sends a DNS query to a recursive DNS server. The DNS client can indicate that it is DNSSEC-aware (DO=1).
2. The recursive DNS server sends a DNS query to the root and top-level domain (TLD) DNS servers. The recursive DNS server can indicate that it is DNSSEC-aware (DO=1).
    a. Retrieved from: http://technet.microsoft.com/en-us/library/jj200221.aspx, 17/12/2015 which consists of a public cryptographic key stored in a resource record which is used to verify signature.
    b. DS: DS (Delegation Signer) creates authentication chains through a secure delegation especially to child zones.
3. The root and TLD servers return a DNS response to the recursive DNS server providing the IP address of the authoritative DNS server for the zone.

Authoritative servers for the parent zone can indicate that the child zone is signed using DNSSEC and include a secure delegation (DS record).

4. The recursive DNS server sends a DNS query to the authoritative DNS server for the zone.

The recursive DNS server can indicate that it is DNSSEC-aware (DO=1) and capable of validating signed resource records (CD=1) to be sent in the response.

5. The authoritative DNS server returns a DNS response to the recursive DNS server, providing the resource record data. The authoritative DNS server can include DNSSEC signatures in the form of RRSIG records in the DNS response, for use in validation.
6. The recursive DNS server returns a DNS response to the DNS client, providing the resource record data. The recursive DNS server can indicate whether or not the DNS response was validated (AD=1) using DNSSEC.

OTHER TECHNIQUES FOR COUNTERMEASURE

Ingress Filtering

It is a means by which service (The ATM Forum 1996) providers can trace the source networks for any relevant traffic. This countermeasure is especially targeted towards spoofing attacks, where the source IP address is used through a proxy or a spoofed IP address. Distributed Denial of Service (DDoS) attacks can also be prevented by using ingress filtering techniques. It can trace the source networks, and also protects itself against spoofed access to networking equipment. There are five different ways of implementing ingress filtering:

1. **Ingress Access Lists:** It crosschecks the source address of every message transmitted over a network against a list of acceptable prefixes and if the packets do not match the filter they are removed.
2. **Strict Reverse Path Forwarding:** It is similar to the concept of using access lists, however, in this case, the access lists, which are used, are of dynamic nature to avoid any issues with duplication using FIB (Forwarding Information Base).
3. **Feasible Reverse Path Forwarding:** It is an extension of Strict Reverse Path Forwarding, including the use of FIB however; it looks up all possible routes invoked with routing protocol specific methods. It is especially targeted towards asymmetric routing or multi-homing of the network.
4. **Loose Reverse Path Forwarding:** It is a type of a "route presence check". It has its similarities to strict RPF, however, it differs in the fact that it checks only if the route exists or not.
5. **Loose Reverse Path Forwarding Ignoring Default Routes:** It is like loose RPF except it does not care for default routes. The router searches the source address in the route table and the packet is preserved if the route is discovered. It varies in the essence as the default routes are excluded.

Unicast Reverse Path Forwarding (RPF)

The priority for any router is to forward the IP packets which it might receive. For this matter, it really does not care about the source IP address, but only cares about the destination IP address. However, this gives an immense opportunity for attackers to attack this particular vulnerability. Therefore, Unicast RPF is employed as a security feature, which can prevent IP spoofing attacks. It consists of a routing table for the source IP address and checks for the IP packet, which it received to verify with the table. If the packet does not match then it is discarded. Unicast RPF works with two modes:

Strict Mode

It means that the router will check for two particular agendas:

1. It searches for the matching entry for the source IP packet in the routing table.
2. It searches for the interface to reach the source

The IP packet is tested for both of these agendas and if the IP packet passes both of them then it is accepted, otherwise, it will be discarded. It is most suitable for IGP routing protocols, which use the shortest path to the source of IP packets concept in which: Interface to reach the source = Interface from where the packets originate.

Loose Mode

Unlike the strict mode, in loose mode, only one agenda is checked: It searches for the matching entry for the source IP packet in the routing table, The IP packet must pass only this particular agenda, and then it is allowed. It is mostly useful when multiple ISPs are involved and asymmetric routing is chosen. However, one disadvantage is the null0 interface related sources are discarded.

Sender Policy Framework (SPF) Protocol

SPF protocol works on the concept of proof of origin and verification method. It basically checks for integrity. SPF is basically a DNS record. The mail server, which is in-charge of the incoming mails, crosschecks the domain of their email address against the list of authorized hosts to send messages of the domain based upon SPF. This process then makes a decision upon whether the email passes the criteria or not, and if it does it is allowed, otherwise it is blocked. SPF protocols are simple enough to use as a user needs to have access to a DNS server and modify a set of parameter to control the email server responses to the email received with SPF and validation. One of the biggest benefits is the prevention of SPAM and forged emails. Forged emails from unauthorized sources, that claim servers that are enabled with SPF verification check false identities. The IP addresses of servers, which generate spams, can be blacklisted using SPF. Therefore, traffic from such servers will be filtered out.

Specific SPF Records

There remains a divide between a pure SPF and TXT record with SPF syntax, where specific SPF varies in terms of TXT record. While most of the modern servers can easily interpret specific SPF, some servers still cannot interpret them. In the case of modern servers, the identity of the message is verified using both records one after the other. First the specific SPF is checked followed by TXT. SPF, which is based upon TXT record in the DNS server, is normally located in the outgoing mail server. The TXT record consists of a list of legitimate IP addresses and host names for a particular domain from which the mail might originate. The incoming mail server can then verify against the TXT record.

MOBILE BANKING

In 2015, The GSMA Intelligence (GSMA, 2015), reported in the report "The Mobile Economy", that there are almost 3.6 billion mobile phone users all around the world which is almost half of the world's population. This reflects the huge increase in users of some sort of mobile phone in the world. However, in developed nations this increase is nominal since the new penetration rate has reached close to saturation. Similarly, the report also suggests that the penetration rate for the adoption of smart phones will also reach 70 to 80 percent ceiling, which is normally considered to be the plateau.

Smart phones use in the developed nation is at its peak; however this increase is mostly accredited to the young users of the smart phones. One of the studies suggested that the reason behind the popularity of mobile phones is the ease of access to information and constant connectivity. There is a difference between the purposes of use of mobile technologies for different age groups. Younger users are concerned with social interaction aspects, while the users who are adult or elderly, the purpose varies from job-related issues to security, safety and personal independence.

The use of Smartphone and mobile technologies, in general, depends upon motivation or purpose of use. This can be either intrinsic or extrinsic motivations and it can also vary in terms of importance, either utilitarian or hedonic. There are several other factors, which are associated with the use of mobile phones, which applies to all users irrespective of age or gender. These factors are accessibility to information, display characteristics, arranging appointments, and safe. One study suggested the use of mobile phones by older people in Finland was highly influenced by security and communication factors. Another study suggested that the purpose of use of mobile phones by the elderly was highly motivated by social integration and independent living.

The sense of security obtained from the ownership of a mobile device is directly correlated to the omnipresent nature of communication medium to contact loved ones at points of distress. However, another added benefit is the ability to remain independent. This increase in the use of secure mobile devices is opening up opportunities for online mobile banking, mobile payment etc. The convenience of doing transactions anywhere and anytime and the increasing level of security on mobile devices are two important factors for this high increase in mobile banking technologies.

Mobile Transaction Procedure

A mobile transaction consists of three entities:

1. A User
2. A Device (mobile)
3. A Transaction operator (Bank, mobile operator, or both)

For security of mobile transactions there are three specific processes:

1. Identification
2. Authentication
3. Secure performance

A mobile user is normally identified in a cellular network through the unique phone number and PIN. The user in a mobile banking environment can be identified by three procedures:

1. What the user has
2. What the user is
3. What the user knows

This means the user can be identified with the device that he possesses or through biometric means of different means ranging from fingerprints, vocal analysis etc. A mobile provider can authenticate the requests from the users through the subscriber information or through various specialized mechanisms such as digital signatures or security protocols such as WTLS (Wireless Transport Layer Security Specification. The transaction process consists of a transaction operator; secure payment protocols such as SET (Secure Electronic Transactions) or iKP (Internet Keyed Payments). The mobile transaction operator can use mobile gateways to support different communication and access control mechanisms. When a transaction request is sent, the device using a private key, which remains unknown to the provider, must sign it digitally. The mutual agreement between a user and the operator sets a public key based upon DSA or RSA. The user must be able to view each individual transaction details for every request using markup languages for signing documents, which are pre-verified with a trusted authority through certificate or signatures.

User authentication is another important aspect. A mobile system is only as strong as its weakest link, which is, of course, a user. With the new smart phones, there are various possibilities to infiltrate devices especially those with open source operating systems. Another important entity involved in online transactions is the merchant. The merchant is same as on the online shopping terms. This merchant account receives payments from a third party such as a financial institution. These transactions normally take place with an establishing a secure communication channel, followed by identification of customer to merchant through subscriber identification or secure signature from customer's device, which is a guarantee of payment.

The security of such a transaction is maintained through a secure communication channel. The user authorizes the transaction request, the payment transaction request from the user's mobile device is matched with the respective offer, and then the payment request contains state information within a cookie or a URL parameter containing MAC (Message Authentication Code) which ensures authenticity through a key which is known only to the operator. After this payment is made, the operator confirms payment to the merchant through a signed authenticated message. The merchant then replies back with a receipt of the confirmation of the transaction.

SMS Banking

SMS banking was initially targeted towards the transmission of non-confidential information through GSM (Global System for Mobile communication) network. The SMS message is quite vulnerable as the messages remain in plaintext and an algorithm called A5 does the encryption. The encryption is done during the transmission process; therefore end-to-end encryption algorithm is not available.

This process has several vulnerabilities, first of which is the sending of a request from the user to the banking server using USSD (Unstructured Supplementary Services Data) string containing the user's PIN code. This sensitive information is sent on the GSM network in plain text. The second vulnerability is the encryption algorithm A5, which is not considered to be secure. Different versions of A5 are used for encryption, A5/1 is the standard used in North America and Europe and A5/3 is used by 3G services under UMTS (Universal Mobile Telecommunications System). Although, 3G services using A5/3 are considered to be secure, however, 3G/UMTS and GSM operate in different frequencies. 3G technologies include two factors:

1. Encryption on the air interface
2. Mutual authentication between the user and the network using HLR (Home Location Register) and USIM (Universal Subscriber Identity Module).

The UMTS frequency can be blocked funneling the user to use GSM. Therefore, the vulnerability still remains in some forms as DCH (Dedicated Channel) starvation attack. DCH is a finite resource divided among the subscribers. When multiple devices connect to a network for DCH services, they soon get depleted and DoS (Denial of Service) attack is instigated.

USSD Banking

USSD (Unstructured Supplementary Services Data) consists of two different modes:

1. USSD 1
2. USSD 2

USSD 1 allows one-way communication to a network while USSD 2 allows for two-way communication between a user and network. End-to-end transaction security is provided throughout the communication layer with a subversion of the identity of the subscriber. However, within the communication layer, the data is not encrypted therefore; the security can be compromised if the encryption is broken.

The sensitive data regarding customers is encrypted and stored on a server in USSD2. The handset only consists of authentication information of customers such as PIN (Personal Identification Number) and banking instructions and the user does not need to provide other credentials such as account number details. However, this system has similar security threat as an ATM card; if the PIN number of the ATM card is known then anyone can use it. Similarly, if the mobile device is stolen then anyone who possesses the device has the device, the SIM card, and the authentication information.

Mobile Banking in GPRS

The early stage of Mobile banking made use of WAP (Wireless Application Protocol) where a user can use the banking service by simply using the URL (Uniform Resource Locator). The WAP enabled mobile banking systems were considered to be secure. However, there is no end to end encryption, between the client and the Gateway, and also between the Gateway and the Server. GPRS (General Packet Radio System) also known as 2.5G is another technology used for data transmission over mobile networks; however, it is inefficient for continuous data transmission. The data transmission services under GPRS are packet-based. GPRS is based upon the penetration rate of GSM. There are 4 major types of data, which have to be secured in a GSM/GPRS network:

- **User Data:** This is voice or non-voice data sent or received by users.
- **Charging Information:** Information used to bill for non-voice services collected from SGSN (Serving GPRS Support Node) and GGSN (Gateway GPRS Support Node).
- **Subscribe Information:** Information about customers recorded in HLR (Home Location Register) and VLR (Visitor Location Register).
- **Technical Information:** Information about GSM/GPRS network architecture and configuration.

Attackers can target any one of these four data types. One particular form of attack is the DoS, which targets GGSN by overloading it with requests for service. This DoS attack causes prevention of other mobile stations to gain access to the network. GSM and GPRS provide three layers of security:

- **Anonymity:** TMSI (Temporary Mobile Subscribe Identities) are used to provide anonymity to subscribers on the cellular network using IMSI (International Mobile Subscriber Identity). The IMSI contains personal subscriber number, home network name and country code of subscription number.
- **Authentication:** In the case of GPRS, authentication is handled by SGSN, through a randomly generated 128-bit number. This number is sent to the mobile station, which generates a 32-bit response using its private authentication key. This key remains unique to the subscriber implemented through SIM (Subscriber Identity Module) and A3 algorithm.
- **User Data Protection:** To provide security against interception and eavesdropping, encryption methods are employed. The random 128-bit number generated in the authentication process and the private key from HLR is combined with the A8 algorithm to produce an encryption key. Therefore, the data transmission process between the GPRS network and mobile station is encrypted using GPRSA5.

SMS AND MOBILE BANKING

This chapter is concerned with one application of mobile banking: Short Message Service (SMS) banking. It could be argued the main long term application of mobile banking will be in the form of the mobile Internet, and the browsing of Internet banking sites on mobile phones, being made possible by an improvement in mobile technologies with browser support for HTML and XHTML, the use of technologies such as Opera's Small Screen Rendering (SSR), the increasing popularity of Cascading Style Sheets (CSS), along with standards such as the W3Cs Mobile Web Best Practices 1.0. This argument would though, overlook the huge popularity of SMS and its advantages such as a flat rate charge, and the fact that customers will not need an expensive Smartphone, I-Phone or Personal Digital Assistants (PDA) style handset.

Short Message Service (SMS), or text messaging, as it is commonly known, is still a tremendous growth area in mobile communications. It is estimated, that worldwide 4.1 trillion (UK) text messages were sent in 2008. The Mobile Data Association (MDA) reports that in 2008 a total of 78.9 billion text messages were sent in the UK alone, 216 million per day, and this was up 22 billion on the annual total in 2007. Research has found that text messaging is most commonly used as an effective one-to-one method of communication between friends, but businesses have also realized that there is huge potential in SMS for carrying out business activities, and for individual communication with customers. It was estimated by market research group Radicati that in 2004, 55% of text messaging was for business use, with much further growth to come. SMS banking services have already been successfully implemented by banks in Asia, the Middle East and South Africa, with both *Push* (automatic) and *Pull* (customer initiated) services offered to customers. At the time of writing the services offered by banks in the UK are limited to *Push* only e.g. the bank sends the customer a weekly account balance, and basic *Pull* services e.g. ordering a new cheque book. The popularity of SMS banking in markets such as India is due in part to the low cost of mobile handsets compared to desktop computers. This may also be true for some socio-economic groups in the UK. The Short Message Service (SMS), a Global System for Mobile Communications (GSM) service, allows the user to send text messages up to 140 bytes. The transmission of a message is carried out by the network operator's Short Message Service Centre (SMSC), which receives the message and routes it to the destination device. A bank can run its own SMSC and in this way generate SMS messages from its own data on its customers' accounts. A weakness with SMS banking is that the messages are not automatically encrypted when they are transmitted. Encryption is possible though, and various software products have been developed for this purpose which would allow businesses to carry out more complicated financial transactions via SMS banking.

There is much current research on handheld devices carried out in the ubiquitous and mobile computing fields. Examples of recent work specifically on text messaging in the UK are studies; Outside the UK there has been considerable research on SMS in society. One major finding of this research is that SMS seems to be a medium favored by the young. Related to this is the rise of 'textish' or 'text-speak', which is a form of abbreviations and has been defined as English with the vowels removed. Textish has had much interest focused on it recently in the UK media. An example of this is the widespread reporting of a 13 year old pupil who wrote an examination essay in textish. Unsurprisingly, the use of textish is most prominent among young people. Other work relevant to this chapter includes research on text entry and mobile phone user interfaces.

Functions of SMS Banking

It is proposed in this chapter that the types of services a bank can offer under the umbrella of SMS banking can be divided into three general functions: transactions, communication/CRM and security. There will be some overlap between these three, but banks could use SMS for each of these purposes separately, or in combination.

Transactions

Ordering a new cheque book or PIN number, requesting a mini statement, transferring money or making a payment, these are all types of banking transactions that could, and are, offered by an SMS service. Examples of such services offering SMS payments are the 'Mobile Wallet' service from T-Mobile and 'm-pay' from Vodafone2. Mobile payments are a form of payment combining elements from other methods of payments such as credit/debit cards, prepaid cards, telephone bills and premium SMS messages (SMS messages that cost a fixed, predetermined amount).

The viability of mobile payments has been generally proved to be acceptable to consumers. In one study, over 80% of participants were willing to accept mobile payment, with 96% giving "privacy of personal data", and 93% giving "simplicity of the method" as reasons for their decision. SMS payment schemes are currently in development by Anam and TR23, and one is already used by the PayPal4 service. The Anam scheme uses SMS text messages to make third party payments, overcoming the necessity for the customer to download software, such as the Monolink solution, to their mobile phone, and also takes advantage of the widespread usage and familiarity with text messaging. In the UK for example, Colchester Borough Council5 has set up a scheme whereby residents can pay their council tax by sending an SMS text 2 Vodafone and T-Mobile launch mobile wallet: TR2, PayPal and Colchester Borough Council.

Communication/CRM

SMS can be used as a one-to-one business to customer communication channel and offers massive potential for customer relationship management (CRM). SMS can be used for marketing of a bank's services and products, confirmation of transactions made by the customer with the bank via another channel (e.g. Internet, telephone banking), confirmation of contact with the bank via another channel, confirmation of appointments, complaints etc. Dealing with customer complaints is an important issue for businesses and for preventing customer switching behavior. Banks need effective channels and procedures for resolving complaints, as this could lead to a customer switching to another bank. An SMS banking service could offer a useful channel for this purpose and there has been some research in this area. There has been little research on mobile CRM so far, but one study focused on an airline using mobile CRM. They argue that customers are not ready for this type of mobile application yet, though they did find that participants who already used the mobile Internet had a more positive attitude. These types of service will generally be of the *Push* type.

Security

SMS can be used as method of adding 2-factor authentication to online transactions, and potentially to telephone transactions. SMS can be used to generate one time pass codes (OTP). An OTP is a password (usually a string of digits) that is valid only for a single online session or transaction that is made available to the customer either by a physical hardware device with a small display that the customer carries; by a Chip and PIN card reader device; or by using an "out-of-band" channel like SMS. To authenticate their transaction a customer must type in the OTP rather than a static password. SMS OTP generation has been implemented in Asia and mobile phone based 2-factor authentication has been researched and proven to work, but there has been little usability research on using SMS as an authentication method.

LOW ADOPTION OF MOBILE BANKING

The advantages of mobile banking appear as convenience, access to banking no matter the location or time, and efficiency. In spite of these advantages some authors have remarked on the slow development of mobile banking. In countries such as Korea, Finland, and Taiwan studies have shown the usage levels of mobile banking are small compared to what would be hoped for. What are the factors preventing large scale adoption of mobile banking? Research has shown that customers worry about how much it will cost along with the security of the service. Though some have argued that security concerns are not a prohibitive issue. The perceived complexity of mobile banking is also argued to be a cause of low usage levels. Recent research has argued that trust is one of the most important factors in the low adoption of mobile banking, and is the factor that most impacts on customer satisfaction with this banking channel. Trust has an impact on level of adoption in all forms of electronic banking, and has been researched extensively. In their survey study on mobile banking in Taiwan, information system (IS) success model with the three quality measures of system quality, information quality and interface design quality. They found that system quality and information quality affected trust more than user satisfaction, and argue that these factors are important in building trust in a mobile banking channel. System quality is defined as the quality manifested in the system's overall performance as measured by a customers' perception. Information quality, including accuracy, is obviously of major importance to any electronic banking channel and will influence customer satisfaction. Lee and Chung argue that interface design quality may be an important factor in building trust, but it is not as important as system quality and information quality. For a bank offering an SMS banking channel, interface design is something that cannot be controlled, as it is dependent on the type of mobile phone the customer owns. With many of the studies described in this chapter there are still questions over how generalizable the findings are, because they are specific to individual countries and cultural factors may play a part. Another major factor in the slow adoption of mobile banking is due to the limitations of mobile devices: tiny screen size, small keypads, reception and network problems and slow connection speeds. Writing a text message is not the easiest thing to do due to the variety of methods of text entry available on mobile devices, and the lack of a standard user interface, or even a standard layout of the keypad. The usability of mobile devices is an important factor in the low adoption of mobile banking. The distinct lack of empirical research evaluating the usability issues surrounding implementing SMS services for either transaction.

Usability

There are a number of definitions as to what usability is, with the first attempted by Miller and based on the concept of *ease of use*, but the most often quoted is the one defined by the ISO as "the efficiency, effectiveness and satisfaction with which specified users can achieve specified goals in a particular environment". As the ISO definition suggests, usability is a multi-dimensional concept.

There are often compromises to be made, trading off different goals to achieve a usable product. An alternative definition from the Usability Professionals Association (UPA) states: "Usability is an approach to product development that incorporates direct user feedback throughout the development cycle in order to reduce costs and create products and tools that meet a user's needs." The first definition gives measurable dimensions with which to evaluate how usable a system/product is, and the second definition suggests a process of how to do this when developing a new system/product, along with the benefits of the approach. Efficiency, effectiveness and satisfaction are independent qualities of the system. Efficiency is concerned with the amount of effort required in usage. It is typically measured as the time taken or the number of clicks to complete a task. Effectiveness is indicative of application robustness and transparency; task completion, accuracy, prevention and easy recovery from errors are typical measures. Satisfaction relates to the degree to which users react positively to their experience whilst completing tasks. There is still not total agreement on what user satisfaction but it can be proposed that it is related to measuring user attitude. It may also concern perceived usefulness, attractiveness and other emotional responses to the system. The ISO definition concentrates on the attributes of efficiency, effectiveness and satisfaction, but there are other components of usability not included in this definition such as learn ability and memo ability. How easy the system is to learn, and how easy is it to remember how to use it again, are very important factor for a casual user. There is also the question of what makes a product useful and successful to a customer or user. All of these definitions suggest that usability has both subjective and objective components that can be measured.

The other usability standard is Human-Centered Design Processes for Interactive Systems. This standard is guidance for anyone who wants to follow a user-centered design process. The standard describes four principle activities that should be followed that will lead to a design that is of high usability. That is, an effective, efficient and satisfying design. The four activities are:

1. Understand and specify the context of use
2. Specify the user and socio-cultural requirements
3. Produce design solutions
4. Evaluate designs against requirements

The standard can be adapted and applied to the design of any product, and the level of effort and the sequence of the activities can vary depending on the type of product being designed. Recently the there has been a trend towards the term 'user experience', and some practitioners distinguish between usability and user experience. Those in agreement with this opinion consider that the concept of usability is too limited, or narrow, to explain the choices people make, and that a user's whole experience with a product should be considered e.g. their thoughts, feelings, perceptions and even their interaction with the company who make/sell the product.

Usability Engineering

The process by which a usable product or computer system is achieved is called usability engineering. The goal of usability engineering is to engineer a quality product that does what it is meant to do, and fulfils a customer's actual needs. It does this by considering the user and following rigorous software engineering methods. The usability engineering process is well established and can apply to all products with a user interface. It follows a cyclic process of design and evaluation followed by redesign and evaluation. The methods used in usability engineering have developed from the fields of ergonomics, human factors and human-computer interaction (HCI), and also use more formal experimental methods favored by psychologists.

Usability engineering can mean different things to different practitioners. At its simplest, usability engineering is the process of applying usability metrics. To others, usability engineering is not just about evaluation, but is involved in the whole development process from the very beginning to the release of the product. Which involves using the documented usability results and feedback from early versions of a system to make changes to subsequent versions? As usability engineering is a "discipline aimed at enhancing the usability of products."

A typical usability engineering lifecycle consists of three basic stages: requirements analysis, design/testing/development and Installation. The lifecycle is a useful tool for integrating usability into software engineering practices. Each of these three stages has many subtasks within, with the middle set of tasks requiring the most attention. Not all projects will require the same level of complexity of lifecycle, and the usability methods employed will be dependent on the timeframe, resources and goals related to that specific project.

SMS and Mobile Banking Usability

Handheld devices offer many challenges for the user. Physically, they are by their very nature, small: they are typically meant to be held in one hand. They have tiny screens, and have small, fiddly keypads and challenging user input methods. They have less memory, CPU power and their connectivity can be slow and unreliable (though this is improving every year). But compared to desktop computers, or even laptops and notebooks, they offer great portability for on the move access to information, and their great power is communication.

Three tests which a device must pass to be considered a handheld device:

1. General operation without cables (except for charging, connecting to desktop).
2. Easily used while being held in one's hands.
3. Addition of applications, or support Internet connectivity.

The popularity of SMS has led to a body of usability research on text entry methods, mobile phones and there has been general research on mobile phone user interfaces (Lee et al., 2006). One explanation for the usability problems encountered by users of mobile devices is a lack of extensive usability evaluations, due to the manufactures' rush to get their products out into the current competitive market place. The tiny key pads on mobile phones have been found to pose usability problems. Thumb size has been shown to cause usability problems with texting. It has been found that older user has usability problems with texting on mobile phones and there has been research into producing mobile phones aimed at the older generation. One study showed that when keys are placed to close together they cause problems for older users. Older user has been found to be passive users of mobile phones, and can find the process of text messaging intimidating.

Much of the previous research in mobile banking has been based on surveys. Usability studies emphasize hands on usage and the collection of performance and qualitative data, and there are few studies observing actual user performance with mobile banking services, and SMS banking in particular.

It is argued that to bring more understanding to the low adoption of mobile banking the usability issues surrounding this channel need to be researched. Usability, along with functionality, both influence real world usage.

It is important that any new product or service is based on user needs and requirements rather than being driven by technology. This is an important issue to consider with SMS and mobile banking. Research related to the usability design issues surrounding online banking are relevant to SMS banking. Customers want electronic banking designs that are secure and have good error prevention functionality. They also want user-friendliness, speed, accuracy and control. If a new banking channel such as SMS banking is to succeed it must also be satisfying to use. This is extremely important as the new channel will be in competition with the other channels offered by the bank, e.g. Internet banking, telephone banking, ATMs and branch based banking. It will be important to compare the new channel of SMS banking to existing channels as a bench mark and to discover any usability issues. Very little of this kind of comparison work has been carried out with mobile banking. Many of the existing SMS banking services use abbreviations.

An example would be that to request an account balance the users sends an abbreviation as ACBAL, but is abbreviations the most usable format? Another issue to consider will be the demographic variables in the customer base, such as age and gender, and how this will affect the accessibility of an SMS banking service. It has already been discussed how text messaging is more popular amongst the young, and the older generation find mobile phones and texting more difficult. It will be important to decide if the marketing of an SMS banking channel should be focused on a specific group of users and how this will affect its cost/benefit ratio to the service provider. These issues will be addressed in this chapter, with the main line of research being concerned with the relationship between usability and the low adoption of SMS banking.

CONCLUSION

In this chapter discussed the concepts related to authentication, online banking processes and other relevant information security issues are discussed further. Literature relevant to these factors is also considered throughout this process of reflection. In next chapter discusses about Web based electronic money for online banking.

REFERENCES

Cheriton & Zwaenepoel. (1985). The V kernel. *Journal ACM Transaction on Computer System, 3*(2), 77-107.

Clark. (1985). The structuring of systems with up calls. *Proceedings of the 10th ACM Symposium on Operating Systems Principles*, 171-180.

Cooper. (1987). Replicated distributed programs. *Proceedings of the 11th ACM Symposium on Operating Systems Principles*, 63–78.

Custer. (1993). *Inside Windows NT*. Microsoft Press.

Das, Sen, & Jayaram. (1996). A dynamic load balancing strategy for channel assignment using selective borrowing in cellular mobile environment. *Proceedings of the 2nd ACM/IEEE International Conference on Mobile Computing and Networking, 3*(5), 333-347.

Deering & Hinden. (1998). *RFC 2460: Internet protocol, version 6 (IPv6) specifications*. RFC.

Draves. (1990). A revised IPC interface. *Proceedings of the USENIX Mach Workshop*.

Paul. (1990). *Cock shot Ps-Algol Implementations: Applications in Persistent Object Oriented Programming*. Prentice Hall.

Persone, Grassi, & Morlupi. (1998). Modelling and evaluation of prefetching policies for context-aware information services. *Proceedings of the 4th ACM/IEEE International Conference on Mobile Computing and Networking*, 55-65.

The ATM Forum. (1996). *ATM User-Network Interface (UNI) Signaling Specification*. ATM Forum/95-1434R11.

Chapter 3
Web–Based Electronic Money for Online Banking

ABSTRACT

Online transaction payment for the purpose of introducing web-based electronic money as an alternative way of online transaction payment, the main areas that cover in this chapter include research of current payment system, limitation of current payment, what are e-money and the current state of electronic money. It will discuss the proposed web-based electronic money as an alternative for online payment and the benefit of web-based e-money. Online payment transaction is a form of a financial exchange that takes place between the buyer and seller facilitated by means of electronic communications for conducting e-commerce and online purchasing. This chapter includes research of current payment system, limitation of current payment, what are e-money and the current state of electronic money.

INTRODUCTION

Mobile banking services are on the complete iterations of current assistance that use the banks presents day structures and infrastructure. Actuality that online banking infrastructure is already in area and reusing existing components is a cost-effective way to build new systems, mobile banking location based banking is as subsequent new release of net banking. For mobile banking we agreed to apply the subsequent infrastructure: 'that kind of execution of economic services within the area of which, the customers make use of

DOI: 10.4018/978-1-5225-2759-6.ch003

mobile communication strategies at the side of mobile devices. In the context of net banking, mobile website banking falls inside that definition. The net banking, internet site accessed is an internet site through a mobile device. A mobile website is a website optimized for viewing on a mobile tool that require authentication in the form of a different device. For financial institution the Random Reader is used handiest for registering for the primary time and, relying at the transaction, during transfers, and for SMS Banking the Digipas is continually used for logging in and for transfers. The Random Reader and Digipas are devices which are used to generate responses to a data obtained from the bank so as to authenticate the consumer.

INFORMATION TECHNOLOGY IN BANKING SECTOR

The banking devices have to protracted manner seeing that independence from nationalization to liberalization. It has witnessed transition from a slow business organization to a extraordinarily proactive and dynamic entity. This transformation has been in large part brought approximately by means of liberalization and economic reforms that allowed banks to discover new commercial enterprise opportunities in place of generating revenues from traditional streams of borrowing and lending. These economic reforms that were initiated in the early Seventies brought in a completely new environment for the banks. The banks are now offering innovative and attractive generation based multi channels to offer their products and services. The process began inside the Seventies, where computers have been introduced as ledger posting machines. Technology has been deployed in variety of again-workplace and consumer-interface supports of banking. In the early Eighties Reserve Bank of India uses a mechanism to speed up the temporary operation in banking sector. An excessive-stage mechanism turned into fashioned to attract up a phased plan for computerization and mechanization inside the banking sector. The consciousness turned into on customer service. For this motive, two models of department have been developed and implemented. The second committee constituted in 1988 a plan for computerization and automation to other areas such as budget control, e mail, BANKNET, SWIFT, ATMs, I-banking and so forth. In the last decade, facts technology has introduced extensive modifications in the banking quarter. It has supplied a possibility to banks for offering different products and services to their clients the usage of today systems. Apart from operations, development in generation has played a vital role within the distribution method of industrial banks. Banks, which have been traditionally counting on important channel i.e. To supply services have now commenced supplying their product and carrier through sort of modern and generation based channels which consist of channels inclusive of Internet Banking', Automated Teller Machines (ATMs)', Mobile Banking', Phone Banking', TV Banking' and so on. All those new channels of distribution are in the area of e-banking or I-banking. Electronic banking has been around for quite a while in the form of automatic teller machines (ATMs) and cell phone transactions. In recent instances, it has been converted by the internet a new channel that has facilitated banking transactions for both clients and banks. As measurement of strategic choices, banks in India have been investing and persisted to make investments extensive amount of finances on laptop and associated technologies anticipating sizeable payoff. According to the Boston Consulting Group (2011), the advanced costs on information technology (IT) for banks on the whole is Rs 6,500 Cr. In keeping with yr, approximately 2.7 in keeping with 100% in their sales is in addition possibly to increase up to Rs 10,000 Cr. Annually in the coming years. Further, Reserve Bank has unique emphasis on generation infusion within the each day operations of banks. The IT Vision Document, 2011-17 of the Reserve Bank design the roadmap for implementation of key IT packages in banking with unique emphasis on faultless release of banking contributions via effective implementation of Business Continuity Management (BCM), Information Security Policy, and Business Process Re-engineering.

DEVELOPMENT OF WEB-BASED ELECTRONIC MONEY FOR ONLINE PAYMENT TRANSACTION

Electronic money or "e-money" (Bakre and Badrinath 1995) is often referred to as a monetary value instrument which is stored electronically on an electronic device such as a chip card or a computer memory. In other words, e-money represents digital money or digital currency. Electronic money is a payment instrument that contains monetary value that has been paid in advance by the user (Baratloo, Chung, Huang, Rangarajan, and Yajnik. Filterfresh 1998). Goods and services can be purchased by users from merchants and the payment can be done through electronic money. The amount will be automatically deducted from their electronic money balance (Badrinath and Gathercast 1998) when they are paying through electronic money. Online payment transaction is a form of a financial exchange that takes place between the buyer and seller facilitated by means of electronic communications for conducting e-commerce and online purchasing (Michael Bender, Davidson, Dong, Drach, Glenning, Jacob, Jia, Kempf, Periakaruppan, Snow, and Wong 1993).

PAYMENT SYSTEM

Payment system is a funds transfer system that facilitates the circulation of money, and includes any instruments and procedures that relate to the system. Payment system is one of the fundamental for the modern economies. Some of the well used payment systems are cash, credit card, debit card, cheque (Birman and Joseph 1987) and electronic money. For this research study purpose, we will divide payment system into conventional payment system i.e. credit card and debit card and alternative payment system i.e. electronic money.

Conventional Payment System

A conventional payment system involves two parties, buyer and seller which a buyer transfers cash or payment information to seller. The payment is settled in the financial institution. For the cash payment, buyer withdraws money from his/her bank account, transfer money to seller and seller deposits the payment to his/her bank account (Birman and Joseph 1987). For non-cash payment, buyer will credit or debit money from his/her account to the seller through credit cards, debit cards or cheques.

Conventional Payment Instruments Adapted to Internet

An overview of the existing payment methods and techniques, which have been developed to adapt the conventional payment instruments for use over the Internet. This research study will focus on conventional payment instruments; credit card and debit card (Birrell and Nelson 1984).

Credit Cards

A credit card is a conventional payment system that entitles its holder to buy goods and services based on the holder's promise to pay for these goods and services. The issuer of the card grants a line of credit to the consumer from which the consumer can borrow money for payment to a merchant or as a cash advance to the user. Each month, a statement will be sent to card holder on indicating the purchases undertaken with the card, outstanding fees and total amount owed. After receiving the statement, the cardholder may dispute any charges that he or she thinks are incorrect. Otherwise, the cardholder must pay a defined minimum proportion of the bill by a due date, or may choose to pay a higher amount up to the entire amount owed. The credit issuer charges interest on the amount owed if the balance is not paid in full. Credit cards are widely used for making payments over internet as they are internationally known to consumers and accepted by merchants (Birrell, Nelson, Owicki, and Wobber 1993).

Debit Card

Debit card provide a convenient way to present the cardholder information needed to debit the card holders bank account. This information is embedded in the magnetic stripe (or chip) on the bank of the card. In some countries, debit cards can be used in internet shops. Internet usage operates similarly to the direct debit system, but offers additional security features for payment owing to the presence of the card. The cardholder readers are many cases provided by card-issuing bank. The use of debit cards for purchases on the internet is still relatively limited.

Limitation of Conventional Payment Instruments

Existing payment systems, such as credit cards and debit card, are inadequate for retail customer digital business from the following general viewpoints (Boyland and Castagna 1997):

Lack of Usability

Existing conventional payment methods, i.e. credit cards and debit cards require consumers to provide a lot of information on web site interfaces before making online payment. E.g. credit card and debit card payments via a web site are not the easiest way to pay, as these require entering extensive amounts of personal data and contact details in a web form.

Lack of Security

Existing conventional payment methods, i.e. credit cards and debit cards has been target of risk and theft abuse. Consumers have to provide personal and account information before making online payment. Even encrypted Internet credit card transactions do not include the owners signature, and anyone with knowledge of the customer's credit card number, expiration date and 3 digit codes can create a payment order. Visa Debit Card and Credit card are an example of an insecure payment system since authentication is based only on "something you know". In order to gain access to their bank account:

1. We need to find out credit card or debit card number, expiry date, and full name.
2. Set up a Visa-enabled merchant
3. Debit the account.

If the victim notices that his account has been debited without his permission, Visa will force the merchant to refund the user. However, if the merchant has disappeared, Visa will refund to the user. This is a cumbersome and expensive process as merchants do not have a reliable way to verify that a credit card or debit card is being used by its registered owner.

Lack of Trust

Consumers would not trust existing payment methods with the long history of fraud, misuse or low reliability. In recent years, there are many reported cases of fraud and thefts on conventional payment systems that has been used as online payment method. Potential customers often mention this risk as the key reason why they do not trust a payment service and therefore do not make Internet purchases.

Lack of Eligibility

Not everyone with money and intention to pay can make use of certain payment methods to make online payment. In the present, majority of e-commerce merchants are adopting credit card and debit card as their online payment method. In reality, not all potential buyers can obtain credit cards and debit card due to credit history limitations, low income or other reasons. In reality, this will hinder on the development of e-commerce.

Lack of Efficiency

Some electronic commerce payments can be too small to be handled by existing payment systems due to high administrative costs included in the processing of payments and transaction. Credit cards and debit cards are too expensive for small payments and unsuitable for small transactions. The minimum fixed fee charged to the retailer for processing a transaction could even surpass the value of the goods sold.

High Usage Costs for Customers and Merchants

Credit card and Debit card are very expensive for consumers as they use expensive infrastructure to assist in the payment process. The growing size of fraud, which amounts to billions dollars per year is intangibly re-financed by users by the higher costs of credit card and debit card services. For online transaction, credit card and debit card are not physically available for inspection, therefore the chance of fraud is higher and charges to merchant will be higher too. Transactions fees are notably higher for internet, between 2.5% and 6% of total sales, depending on the chargeback history of the merchant. For internet transaction, payments must be approved in real time by the card issuing bank. Online authorization will help to prevent fraud but will increase handling costs for all parties involved. Average, online authorization process takes about 6 to 90 seconds. There is a risk that consumers will reject the transaction before completion due to unacceptable queuing for online authorization. If the payment fails, the merchant must either reject the payment or accept a much higher chargeback risk. In addition, credit card and debit card bills are sent in a paper form to customers by post, and the bills are mostly settled by posting paper documents, which makes the whole cycle rather expensive.

ELECTRONIC MONEY

Electronic money is a payment instrument (Carter and Crovella 1997) that contains monetary value that has been paid in advance by the user. Goods and services can be purchased by users from merchants and the payment can be done through electronic money. The amount will be automatically deducted from their electronic money balance when they are paying through electronic money. Online payment transaction is a form of a financial exchange that takes place between the buyer and seller facilitated by means of electronic communications for conducting e-commerce and online purchasing.

In general, electronic money products are "stored-value" or "prepaid" products in which a record of the funds or "value" available to a consumer is stored on an electronic device in the consumer's possession. The electronic value is reduced whenever the consumer uses the device to make purchases and intended to be used as a multipurpose means of payment. Electronic money allows consumers to use electronic means of communication to make payment. Banks may participate in electronic money schemes as issuers or distribute electronic money issued by other entities; redeeming and maintaining electronic money transactions for merchants; handling the processing, clearing, and settlement of electronic money transactions.

CURRENT STATE OF ELECTRONIC MONEY

The continuous change from paper-based payments to electronic form is obvious from the rising trend in the number of electronic payment transactions recorded in 2008. The motivating force for this upward trend is the consumer demand for fast, convenient and secure transactions, as well as the merchants'' efforts in improving business processes and lowering costs. Payment cards are still the most popular e-payment mode used with electronic money (e-money) recording the highest number of transactions and credit cards leading the way in terms of amount spent. At the same time, the e-money industry continued to gain reputation as an alternative payment instrument for micro payments, representing more than half of non-cash transactions performed in the economy.

CONCLUSION

This chapter includes research of current payment system, limitation of current payment, what are e-money and the current state of electronic money. It will discuss the web-based electronic money as an alternative for online payment and the benefit of web-based e-money. Next chapter discusses about Mobile transaction processing system in real world.

REFERENCES

Badrinath & Sudame. (1998). *An efficient mechanism for multi-point to point aggregation in IP networks.* Technical Report DCS-TR-362. Rutgers University.

Bakre & Badrinath. (1995). I-TCP: Indirect TCP for mobile hosts. *Proceedings of the 15th International Conference on Distributed Computing Systems.*

Baratloo, Chung, Huang, Rangarajan, & Yajnik. (1998). Filterfresh: Transparent hot replication of Java RMI servers. *Proceedings of the USENIX Conference on Object-Oriented Technologies (COOTS).*

Bender, Davidson, Dong, Drach, Glenning, Jacob, … Wong. (1993). Unix for nomads: Making Unix support mobile computing. *Proceedings of the USENIX Symposium on Mobile & Location-Independent Computing.*

Birman & Joseph. (1987a). Exploiting virtual synchrony in distributed systems. *Proceedings of the 11th ACM Symposium on Operating Systems Principles.*

Birman & Joseph. (1987b). Reliable communication in the presence of failures. *ACM Transactions on Computer Systems, 5*(1), 47–76.

Birrell & Nelson. (1984). Implementing remote procedure calls. *ACM Transactions on Computer Systems, 2*, 39–59.

Birrell, Nelson, Owicki, & Wobber. (1993). Network objects. *Proceedings of the 14th ACM Symposium on Operating Systems Principles.*

Boyland & Castagna. (1997). Parasitic methods: An implementation of multi-methods for Java. *Proceedings of OOPSLA '97: Object-Oriented Programming Systems, Languages and Applications*, 66-76.

Carter & Crovella. (1997). Server selection using dynamic path characterization in wide-area networks. *Proceedings of the IEEE Conference on Computer Communications (INFOCOM 97).*

Chapter 4
Mobile Transaction Processing System

ABSTRACT

In this chapter, we first revisit the basic concepts of database transactions, and discuss how these concepts are achieved in practical systems. Next, we briefly go through the architecture of transaction processing systems in the centralized and the distributed environments. This chapter we have reviewed the basic concepts of database systems and database transactions, and discussed the architecture of transaction processing systems in distributed environments. We will shift our focus to transactions and transaction processing in mobile environments, which possess some unique characteristics such as the mobility of mobile computing hosts, the limitations of wireless communications and the resource constraints of mobile computing devices.

INTRODUCTION

Mobile banking (also known as m-banking) is the time period used for performing banking transactions or accessing financial services via a cellular tool along with a mobile cell phone. It has revolutionized the banking enterprise with new business styles to offer handy self-service banking options to their clients. With cellular banking, a client may be sitting at a far flung region, however so long as the consumer has a cell telephone with network connectivity, the consumer can get admission to his/her account anytime, anywhere. For a consumer to apply mobile banking, the bank calls

DOI: 10.4018/978-1-5225-2759-6.ch004

the consumer to sign up for the services. During registration, the consumer gets (or provides) a four or five digit Personal Identification Number (PIN) as a password. To get entry to the provider, the purchaser is required to get into the perfect combination of his/her identification (generally the account number or the cellular variety) and the registered PIN to authenticate. Yet, this mechanism is unsatisfactory. The use of a textual content-based password calls for a change-off among safety and exorability; the exchange-off arises from the issue of human reminiscence, and, as an end result, passwords are effortlessly forgotten.

To keep away from the threat of forgetting passwords, users often adopt insecure behaviors, which include writing down their passwords and storing them in an in-secured place or disclosing their passwords to perceived trusted parties. Users adopt such insecure behaviors due to the fact that they lack protection cognizance; and that they often construct their personal faulty model of viable safety threats. As a result, users neglect the importance of training correct security conduct. It carried out a look at to recognize the factors influencing peoples 'safety behaviors. Their findings display that people's misbehaviors are often because of negligence and lack of understanding. To pressure customers to adopt the precise conduct, they advocate companies to use the concern appeals method (a method of persuasion by way of frightening human beings to confirm with a specific message and also describing its bad effects if the message results in, not obeyed) to influence customers in schooling and on-line help. Although presenting protection schooling should boom users 'protection focus, but, it does not improve the usability of a safety device. At the same time, this method increases the burden on the customers; rather than educating customers about system safety. It is greater vital to build a gadget with usable safety. In the hunt, in the direction of a usable security answer for cellular banking, in this dissertation, we're specifically interested in exploring the usability of password systems for authentication of the usage of cell telephones.

DATABASE AND TRANSACTION CONCEPTS

A database is a collection of data items that is gathered over a period of time, and safety stored for further examination or analysis. A database is usually accompanied by a data structure and a set of constraint rules that specify what information (Maassen, Nieuwpoort, Veldema, Bal and Plaat 1999) a data item represents. For example, in an employee database, the employee age is an integer number and must be greater than eighteen and less than sixty five. A database state is a collection of all the stored data values of all the data items in the database at a specific time. A consistent state of a database is a database state in which all the data values fulfill all the constraint rules of the database. A set of operations is usually provided to support users in retrieving or modifying data items in the database. These provided operations can be simple, for example read and write operations, or more complex operations, for example deletion or modification operations. To assist users to perform much more complex operations rather than reading from and writing to the database, a piece of specialized software called a database management system (DBMS) is accommodated to the database. In general, a DBMS not only provides an easy-to-use and friendly interface to users for accessing and manipulating the database, but also manages all the database operations. In addition, the DBMS also protects the database from unauthorized users.

Database Transactions

Users can interact with the database by one or many database operations. The database operations can be gathered together to form a unit of execution program that is called a transaction. In other words, a transaction is a logical execution unit of database operations. A transaction transforms the database from one consistent state to another consistent state.

Transactional Programming Model

A transaction program starts from an initial consistent state of the database by invoking a Begin_transaction method call. After that, one or a set of database operations of the transaction program are executed. When these database operations are completed, i.e., a new consistent database state is established as designed, the transaction program saves this new consistent state into the database by calling the Commit_transaction method. The Commit_transaction call ensures that all the database operations of the transaction program are successfully executed and the results of the transaction are safely saved in the database. If there is any error during the execution of the transaction program, the initial consistent state of the database is re-established by the Abort_transaction call. The Abort_transaction call indicates that the execution of the transaction program has failed and this execution does not have any effect on the initial consistent state of the database. The transaction is said to be committed if it has successfully executed the Commit_transaction call, otherwise it is aborted. A transaction is called a read-only transaction if all of its database operations do not alter any database state.

The ACID Properties

In a database system, there may be a large number of transactions that are executed concurrently, i.e., the shared data items in the database are read and possible written by many transactions at the same time. Each transaction must ensure that it always preserves the consistency of the database system. In order to retain and to protect the consistency of the database system, transactions will have the following ACID (Martin, Bergan, and Russ 1988) (Atomicity, Consistency, Isolation, and Durability) properties:

Atomicity

Either all database operations of a transaction program are successfully and completely executed, or none of the database operation of this transaction program is executed.

Consistency

A transaction must always preserve and protect the consistency of the database, i.e., it transforms the database from one consistent state to another. In other words, the result of a transaction that has committed fulfills the constraints of the database system.

Isolation

An on-going transaction must not interfere with other concurrent transactions, or be able to view intermediate results of other concurrent transactions.

Durability

The result of a transaction that has successfully committed is permanent in the database. The consistent state of the database is always survived despite any type of failures.

The ACID properties of a transaction ensure that:

1. A transaction always keep the database in a consistent state
2. A transaction does not disturb other transactions during their concurrent execution processes
3. The consistent state of the database system that is established by a committed transaction withstands software or hardware failures. In order to achieve the ACID properties, normally, two different sets of protocols named concurrency control protocols and recovery protocols are needed.

Concurrency Control of Transactions

We discuss the problems that can occur in a database system in which there are many transactions being executed concurrently. In other words, we answer the question of why there is a need of concurrency control in the database system. We also review different techniques that ensure the correctness of transaction execution. To illustrate and to simplify the analyses without losing generality, we assume that each transaction possesses the following characteristics:

1. Transaction Ti starts by a Begin_transaction call that is denoted by Bi.
2. A database operation Opi(X) on a data item X is either a read operation Ri(X) or a write operation Wi(X). In general, more complex operations on a database system can be modeled via read and write operations.
3. Transaction Ti ends by either a Commit_transaction call denoted by Ci, or an Abort_transaction call denoted by Ai.

Some typical problems which are caused by the concurrent execution of transactions are: lost update, dirty read, and unrepeatable read. First, the lost update occurs when two transactions T1 and T2 try to write the same data item X. In the figure, transaction T2 overwrites the value of data item X that was prior written by transaction T1. The dirty read occurs when transaction T2 reads the value of data item X that is written by transaction T1 before the transaction T1 commits. If the transaction T1 aborts, the transaction T2 has been operating on an invalid data value. Finally, the unrepeatable read happens if a transaction executes the same read operation at different times, and obtains different data values. The read operations of transaction T2 return two different values of X: before and after the write operation of transaction T1. The concurrency problems can be solved if the DBMS can schedule these database operations of transactions in an execution order in which no transaction interferes with other, i.e., fulfills the isolation property of transactions. The execution order that sequentially contains all the database operations of all concurrent transactions is called the schedule or history of transactions. The order of database operations of one transaction must be retained in the schedule of all transactions. A schedule is a serial schedule if, for any pair of transactions, all the database operations of one transaction follow all the database operations of another transaction. In other words, the isolation property of transactions is ensured in a serial schedule.

The main disadvantage of the serial schedule is that transactions must be executed serially, i.e., the concurrent execution of transactions does not exist in a serial schedule. This may decrease the performance of the database system. To deal with this drawback, the concept of serializable schedule is normally used. A schedule is serializable if it is equivalent to a serial schedule. The remaining question is how to determine if a schedule is a serializable schedule. In other words, we need to clarify the "equivalent" term. Two examples of the equivalent serializability are: conflict serializability and view serializability.

Conflict Serializability

The conflict serializability is based on the concepts of conflicting operations. The idea behind the conflicting operations is that: for two sequentially executed operations Op1 and Op2 that belong to two transactions T1 and T2 (McCann and Roman 1999), respectively, if their order is interchanged, i.e., Op2 Op1, the results of at least one of the involved transactions will possibly be changed. In other words, two database operations that belong to two different transactions are conflicted if they access the same data item in the database and at least one of them is a write operation. Two consecutive operations, which are not in conflict, can be swapped or interchanged in a schedule without any effect on the transaction behavior. Two schedules are said to be conflict equivalent if one can be turned into another by swapping the pairs of non-conflict operations. A schedule is conflict serializable if it is conflict equivalent to a serial schedule.

Verify Conflict Serializable

A schedule S can be validated if it is conflict serializable by analyzing a serialization graph. A serialization graph (SG) is a directed graph that is constructed in two steps as follows:

1. Each node labeled Ti in the SG represents an equivalent transaction Ti in the schedule S.
2. Any pair of operations, Opi and Opj, that are conflict in the schedule S, and Opi precedes Opj, add an edge from Ti to Tj in the SG.

The schedule S is conflict serializable if the constructed SG has no cycles. The serialization graphs of schedules CS1, CS2 and non-CS3 are constructed. For schedules CS1 and CS2, the corresponding SG do not contain any cycle, i.e., the schedules are conflict serializable. On the other hand, the SG of the schedule non-CS3 does contain a cycle T1→T2→T1, i.e., it is not conflict serializable.

View Serializability

View serializablity is a weaker condition that guarantees that a schedule is serializable. Two schedules S1 and S2 are said to be view equivalent if the following conditions hold:

1. Any read operation in either schedule returns the same data value
2. If a write operation Wi(X) is the last operation on data item X in S1, Wi(X) must also be the last operation on X in S2. Thus, the view equivalent conditions ensure that all the transactions read the same data values, and the final database states are identical. If a schedule is view equivalent to a serial schedule, it is said to be view serializable. Figure 2.6 illustrates a view serializable schedule. The serial schedule S1 presents the sequential order schedule of transactions T1, T2, and T3. The schedule VS2 is not a conflict serializable schedule because of conflict operation pairs ((W1(X), R2(X)) and ((W2(Y), W1(Y)). However, the schedule VS2 is a view serializable schedule because: (1) all the read operations R1(Y), R2(X) and R3(X) return the same data values

of data items Y and X as in the serial schedule S1; and (2) all the write operations W1(X) and W3(Y) are the last write operations on the data items X and Y as in the serial schedule S1. The main disadvantage of view serializability is that, verifying view serializable schedule problem has been shown to be a NP-complete problem, i.e., it is not likely that a polynomial time algorithm for this problem will be found.

Concurrency Control Protocols

To assure that a schedule S is serial equivalent, the database system must keep track of conflict operations in the schedule S, constructs the SG of the schedule S, and checks for a cycle in the constructed SG. This process repeats every time when a new database operation arrives to the database system, and requires a lot of computing resources and processing time. Due to the overhead of checking serialization graphs, one normally requires that a completion of the execution schedule of all committed transactions is available before the verifying algorithm can be carried out. This is not true in real-world transaction processing systems where transactions are dynamically and continuously submitted to the transaction processing system. Concurrency control protocols, in fact, do not check for serializability, but are used to ensure that a sequence of executable database operations submitted (Mogul, Rashid, and Accetta 1987) from on-going transactions can form a serializable schedule. There are two main approaches for concurrency control protocols: pessimistic (also called guard-before) and optimistic (also called guard-after). For the pessimistic approach, a database operation is checked if it could cause a non-serializable schedule before it is executed. The database operation is rejected, i.e., the transaction is aborted, if it may potentially lead a schedule into a non-serializable schedule. For the optimistic approach, the submitted database operation is immediately executed as if there is no conflict between this database operation and database operations of other transactions. When a transaction begins to commit, a certification process, in which the transaction will be validated against other transactions, is carried out. If none of the database operations of this transaction breaks the serializability, the transaction is allowed to commit, otherwise the transaction is aborted.

Locking and timestamp ordering protocols are two common concurrency control protocols that are mostly used in the pessimistic approach. Concurrency control by the locking protocol requires that a transaction must request an appropriate lock on a data item before its database operation can be accepted for executing. In other words, a lock plays a role as an execution license for the database operation. One usually applies two types of lock: shared (read) and exclusive (write). A shared lock can be granted too many transactions at the same time, while an exclusive lock can only be assigned to one transaction at a time. Serializability among transactions can be guaranteed by a 2-phase locking (2PL) protocol. The 2PL protocol requires that a transaction must obtain all its locks (in growing phase) before it can release any lock. Strict 2PL is a locking protocol that only allows a transaction to release exclusive locks after it has committed or aborted.

Concurrency control by using timestamp ordering guarantees serializability among transactions based on the following time quantities:

1. The starting time or timestamp of each transaction TS
2. The read and write timestamp values for each data item X, denoted by Read_TS (X) and Write_TS(X) respectively. These read or write timestamp values correspond to the timestamp value of the latest transaction that successfully reads or writes the data item X. A timestamp can be a computer system clock or any logical counter maintained by the database system. When a transaction submits a database operation on a data item X, the timestamp TS of the transaction will be checked against the current read Read_TS(X) and write Write_TS(X) timestamp values of the data item. The outcome of this timestamp checking procedure is either the database system accepts the submitted database operation and the new timestamp value is updated for X, or the transaction is aborted.

There are several methods to carry out the certification process of a transaction, for example the serialization graph testing (SGT) or the validation. The SGT method dynamically builds a serialization graph SG between transactions when a conflicting operation is carried out, when a transaction Ti requests to commit, the SGT method checks if the transaction Ti belongs to a cycle of the SG. If it does, the transaction Ti is aborted; otherwise the transaction Ti passes the certification procedure and will be allowed to commit. The validation method is based on the concepts of conflicting operations to ensure that the scheduling of a transaction Ti is serializable in relation to all other overlapping transactions Tj, which have not committed when the transaction Ti begins.

Every concurrency control protocol has disadvantages. Transactions in a database system that uses locking protocols can suffer from deadlocks or long blocking periods. Timestamp ordering protocols could have decreased the performance of the transaction processing system if there is a high conflict among transactions, i.e., many transactions must abort or roll back. For guard-after approach, works that have been done and system resources might be wasted if transactions are aborted. Concurrency control in a database system can apply either one or a combination of these concurrency control protocols.

Recovery Concepts

The objective of recovery protocols is to enforce the atomicity and durability properties of transactions. The atomicity property requires that either all or none of the database operations of a transaction is carried out. The durability property refers that the results of committed transactions, i.e., consistent database states, survive any kind of failure. In this section, we first study different types of failures that could happen in a database system. Later, we review different recovery techniques that allow the database system to recover from failures.

Type of Failures

Normally, databases are stored on non-volatile media systems like magnetic or optical disks, and are further backed-up by one or more safe storage systems. During the execution of transactions, data items are loaded and temporarily stored in computer memory that is volatile storage.

There are two main types of failures of a database system: catastrophic and no catastrophic. A catastrophic failure happens when there is a breakdown in data storage systems, for example a hard disk crashes. A catastrophic failure can be recovered if there is a sufficient database system backup. Non-catastrophic failures do not affect the non-volatile database storage system, i.e., only data in the volatile storage such as memory is lost. The non-catastrophic failures include transaction and computer system malfunctions. Failures of transactions might be caused by logical faults of data or transaction programs or by the database system. Computer system malfunctions could be caused by errors in the operating systems or applications. A recovery support system will keep track of and record the progress of the execution of transactions by periodically writing important information like data modifications, commitments or abortions of transactions to a logbook, which is stored in the non-volatile storage system. These log records will be used to re-establish a consistent database state if any failure occurs.

Undo vs. Redo Approaches

There are two main recovery techniques that are undo and redo. These two approaches support the database systems to reconstruct consistent database states when there is any failure in the database systems. However, they are different in logging strategies. The undo logging strategy records in the non-volatile logs the former consistent database states before these database states are changed by a transaction. The redo logging writes to the non-volatile logs the new consistent database states that the database systems will have after the updated transaction commits.

The undo technique supports the database systems to reconstruct the previous consistent database states when a transaction fails. The database system behaves as if none database operation of the aborted transaction has been executed. In other words, the undo technique is used to clean up the presence of data values of uncommitted transactions in the database system. For the undo approach, the new database states must be written to the database systems after the undo logs have been written to the non-volatile storage. Redo technique endorses the database system to re-produce the database states that are the results of successfully committed transactions. The redo approach, therefore, will ignore any uncompleted transaction. Before the new data values are written to the database systems, all the redo log records must be written to the non-volatile storage. A recovery support system can combine (which is also the normal case) both undo and redo approaches so that it can decrease the work lost by failures. If transaction T1 aborts after it has modified the value of data item Y, the recovery system can re-establish the initial database states by two logging records <T1,X,10> and <T1,Y,20>. For the redo approach, if a failure occurs after transaction T1 has committed, the database system will re-produce the committed values of transaction T1 based on two logging records <T1,X,20> and <T1,Y,10>. If a new failure happens when the database system is being recovered from previous failures, the recovery procedure has to be able to restart as many times as needed. This feature is called idempotent, i.e., the results of the re-executed recovery procedure are independent of the number of times that they are repeatedly executed.

Recoverability and Cascading Abort of Transactions

When a transaction is aborted, its effect on the database system will be rolled back. If a transaction commits, its results are permanent by the durability property. In other words, a committed transaction does not rollback. A schedule S is said to be recoverable if no transaction T in S commits (Montenegro and Drach 1995) until all transactions T' that have updated data items that T reads have committed or aborted. A serial schedule is, therefore, always recoverable. Note that a serializable schedule does not forbid a transaction Ti to read from a data item X that is modified by an uncommitted transaction Tj. Recovery techniques make no attempt to support the serializability of transactions. Schedule S3 is a recoverable schedule because the transaction T2 that reads new value of data item X modified by the transaction T1 commits after the transaction T1 has committed. Schedule S4 is a serializable but non-recoverable schedule because transaction T2 commits before T1 commits.

In a recoverable schedule S, a transaction Ti reads data values that are written by an uncommitted transaction Tj, if transaction Tj aborts, Ti must also abort. The abortion of transaction Ti could subsequently cause other transaction Tk to abort if the transaction Tk has been reading data values that are modified by the transaction Ti. This abortion could recursively happen to many other transactions. This phenomenon is called cascading abort. Unfortunately, recoverable schedule does not prevent the cascading abort problem. Therefore, a stronger condition that only allows a transaction to read data values, which are modified by committed transactions, is needed. An avoid cascading abort schedule only allows a transaction to read data values that are written by a committed transaction. Furthermore, a strict schedule only allows a transaction to read or write data items that are modified by committed transactions.

First discuss the basic and essential components of a transaction processing system that manages the execution of transactions on a transaction-oriented database system. Later, we review the architecture of distributed transaction processing systems.

Essential Components of a Transaction Processing System

A transaction processing system plays a role as a mediator that accepts transaction requests from users, dispatches these requests to the database system, coordinates the execution of the involved transactions, and forwards transaction results to the original acquirers. The common programming model for a transaction-oriented database system is the client-server model. Users or clients interact with the database system by submitting their transaction processes that consist of one or many database operations to the transaction processing system. The transaction processing system will coordinate and manage the execution of these transaction processes by subsequently sending these database operations to the database system. The database system will carry out the actual execution of the submitted database operations. Finally, the transaction results that reflect the consistent states of the database system are returned to the clients.

To protect the integrity constraint of the database system, the transaction processing system must ensure that the ACID properties of transactions are fulfilled. In order to achieve this, a set of essential components that includes a transaction manager, a scheduling manger and a log manger are deployed. Additional components such as communication manger or other resource managers can also be employed by the transaction processing system. However, in this section, we will focus our discussion on the three essential components.

The role of each transaction processing component is described as follows:

Transaction Manager

The role of the transaction manager is to orchestrate the execution of transactions. Via the help of the scheduling and log managers (explained below), the transaction manger takes care of all important operations of transactions such as begin, read, write, commit, and abort (or rollback). If the execution of a transaction is distributed to many different resource managers, the transaction manager will act as the coordinator of the involved participants.

Scheduling Manager

The scheduling manger manages the order of execution of the database operations. Usually, the scheduling manager makes use of concurrency control protocols, for example locking or timestamp protocols, in order to control the execution of transactions. Thus, the scheduling manger supports the isolation and consistency properties of transactions. Based on the applied concurrency control protocol, the scheduling manager will determine an execution order in which the submitted database operations will be carried out. For example, if a locking protocol is used, the scheduling manager will decide whether a lock request will be granted to the acquired transaction, or if a timestamp protocol is applied, the scheduling manager will assess if a submitted operation will be allowed to be carried out.

Log Manager

The role of the log manager is to support the database system to recover from failures. The log manager keeps track of the changes of the database states by recording the history of transaction execution. Depending on the deployed recovery strategies, for example undo and/or redo, the log manger will record necessary information in a non-volatile logbook. The log manager ensures the atomicity and the durability properties of transactions. The cooperation among the transaction manager, the scheduling manger and the log manager will assure that the ACID properties of transactions in a transaction-oriented database system will be fulfilled.

Distributed Transaction Processing Systems

In the previous section, we have discussed the essential components of a transaction processing system where data is stored in one database system. In this section, we will consider a distributed database system where data is distributed among different computers. A distributed transaction processing system is a collection of sites or nodes that are connected by communication networks.

The communication networks are usually reliable and high speed wired networks, like LANs or WANs. At each node in a distributed system, there is a local database management system and a local transaction processing system (TPS) that operates semi independently and semi-autonomously. An execution of a transaction in a distributed database system may have to spread to be processed at many sites. The transaction managers at different sites in a distributed transaction system cooperate for managing the transaction execution processes.

Transactions in a distributed system can be categorized into two classes: local transaction and global transaction. Consequently, there are two types of transaction manager in a distributed transaction processing system: local transaction manager and global transaction manager. Local transactions are submitted directly to local transaction managers. Local transactions only access data at one database system at one site, and are managed by the local transaction manager. On the other hand, global transactions are submitted via the global transaction manager. A global transaction can be decomposed into a set of sub-transactions; each of which will be submitted and executed as a local transaction at a local database system. Therefore, the execution of a global transaction can involve accessing data at many sites, and be under control of many local transaction managers. A successful global transaction must meet both the integrity constraints of local databases and the global constraints of the distributed database system.

Some of the potential advantages of the distributed transaction processing system are:

1. Higher throughput for transaction processing
2. Higher availability than the centralized transaction processing system. However, the distributed transaction processing system also introduces many challenging issues, for example disconnections in communication between computing sites or concurrency control across computing sites. These problems could cause data inconsistent among database systems, and abort on-going transactions. Consequently, more complicated concurrency control protocols or transaction commitment protocols are needed, for example distributed 2-phase locking and 2-phase commit protocols. Moreover, the heterogeneous characteristic of the distributed system must also be taken into consideration for example different database systems or operating systems.

Mobile Transaction Processing Systems

This chapter focuses on the main topic of this chapter (Mummert, Ebling, and Satyanarayanan 1995): mobile transaction processing systems. The main objective of this chapter is to identify a set of requirements that must be fulfilled by a mobile transaction processing system in order to efficiently support transaction processing in mobile environments.

Unlike distributed environments, transaction processing in mobile environments must take into account three new challenging characteristics of mobile environment – that are: the mobility of mobile computing hosts, the limitation of wireless communications and the resource constraints of mobile computing devices. These three challenging characteristics have a strong impact on the processing of transactions in terms of concurrency control, data availability, and recovery strategies. Because of these unique characteristics of the mobile environments, the standard transaction ACID properties can be too strict to be applied in mobile environments. In other words, we need to define a set of requirements that broadens these properties in the context of the mobile environments.

Characteristics of Mobile Environments

Discuss the characteristics of the mobile environments that could have strong impact on mobile transactions in terms of transaction specification and transaction processing. There are other important issues like authentication and security; however, they are not in the scope of this chapter. The main characteristics of the mobile environments that are addressed in this section include: the mobility of mobile computing hosts, the limitation of wireless communications and the resource constraints of mobile computing devices. In this chapter, we will use the mobile transaction terminology for specifying transactions in mobile environments.

Mobile Hosts

Mobility is the main characteristic that distinguishes the mobile environments from the traditional distributed environments. In traditional distributed environments, computers are stationary hosts. In mobile environments, mobile computers are continuously moving from one geographical location to another.

The features of the mobility characteristic are discussed as follows:

Real-Time Movement

The mobility of the mobile host is a real-time movement. Therefore, it is affected by many environment conditions. For example, the preplanned travel route of a mobile host can be changed because of traffic jams or weather conditions. If there is a mobile task whose operations depend on the travel route of the mobile host, these operations can become invalid, or extra support is required. For example, a new route-map directory must be downloaded into the mobile host if the travel course is changed. Moreover, the movement of the mobile host can also depend on the objective of the mobile task. For example, an ambulance car wants to arrive at the accident scene by selecting the shortest route with fastest allowing speed, a bus must follow a strict time table on a bus-route, while a postman only wants to travel through each road once. During the movement, the mobile host can stop at some locations for some periods; therefore, the mobility of the mobile host includes both movement and non-movement intervals.

Change of Locations

The location of a mobile host changes dynamically and frequently in accordance with the speed and the direction of the movement. The faster the mobile host moves, the more frequently the location changes. The objective of mobile tasks can also specify the locations at which the mobile host must be, in order to carry out the mobile tasks. For example, a computer technician must come to customer locations to fix computer problems. A mobile support system must provide the utilities to manage the locations of mobile hosts (this demand is not needed in a distributed environment). Changes of locations can cause changes in the operating environments of the mobile hosts, for example network addresses, communication protocols, mobile services, or location dependent data.

The mobility of mobile hosts will have strong impact on the execution of transactions. The real-time movement of mobile hosts could pose timing constraints on the execution schedule of transactions. Furthermore, if mobile hosts change their locations frequently, additional time is required to reconfigure transaction application processes to the new environment conditions. Therefore, additional support is required for mobile transaction processing systems to cope with these challenges.

Wireless Networks

Mobile hosts communicate to other hosts via wireless networks. Compared to wired networks, wireless networks are characterized by: lower bandwidth, unstable, disconnections, and ad-hoc connectivity. The characteristics of the wireless networks are described as follows:

Lower Bandwidth

The bandwidth of a wireless network is lower than a wired network. The wireless network does not have the capacity as the wired network. For example, a wireless network has bandwidth in the order of 10Kbps or a wireless local area network (WLAN) has bandwidth of 10Mbps; while gigabits (Gbps) are common in wired LAN. Therefore, it can take longer time for a mobile host to transfer the same amount of information via the wireless network than the wired network. Consequently, the wireless network introduces more overhead in transaction processing.

Unstable Networks

A wireless network has high error-rates, and the bandwidth of a wireless network is variable. Due to errors during data transmission, the same data packages are required to re-transmit many times, thus, extra overhead in communication and higher cost. Due to the varying bandwidth, it is hard to estimate the time required to completely transmit a data package from/to a mobile host. These problems will affect the data availability at the mobile hosts. As a result, the execution schedule of transactions at the mobile hosts can be delayed or aborted.

Disconnections

Wireless networks pose disconnection problems. Disconnections in communication can interrupt or delay the execution processes of transactions. More seriously, on-going transactions could be aborted due to a disconnection. The disconnection in communication is categorized into two types: disconnection period and disconnection rate.

Disconnection Period

The disconnection period indicates how long a mobile host is disconnected. While being disconnected, the mobile host will not be able to communicate to other hosts for sharing of data. If the mobile host holds vital shared data, it can block transaction processes on other hosts. Furthermore, the duration of a disconnected period of a mobile host is not always as planned, i.e., it can be longer than expected. The mobile transaction processing system must be able to continuously support transaction processing while the mobile host is being disconnected from the database servers by caching the needed data beforehand.

Disconnection Rate

The disconnection rate indicates how often the wireless communication is interrupted within a predefined unit of time. The execution of transactions on a mobile host can be affected when an interruption occurs. The more interruptions the many transactions are aborted or rollback. If the transactions on the mobile host are carrying out collaborative operations with other transactions on other mobile hosts, these collaborative activities can be suspended or aborted. To cope with this problem, the mobile transaction processing system must be able to support the mobile transactions to resume or recover from previous interrupted points.

Ad-Hoc Communication

The wireless network technologies introduce a new way to support direct and nearby communications among mobile hosts, also called any-to any or mobile peer-to-peer communication. For example, two mobile hosts can directly share information with the support of Bluetooth or infra-red technologies. The characteristics of this peer-to-peer communication are: unstructured (i.e., ad-hoc), short-range, and mobility dependent.

Computing Devices

There are many types of mobile computing devices such as mobile phones, laptop computers, or personal digital assistants (PDAs). Mobile devices are subject to be smaller and lighter than stationary computers. Consequently, mobile computers have limited energy supply, less storage capacity, and limited functionality compared to stationary computers. Furthermore, the mobile computers are easily damaged, i.e., less reliable. The characteristics of mobile computing devices are elaborated as follows:

Limited Energy Supply

The operation of mobile computers heavily depends on the electrical power of batteries. This limited power supply is one of the major disadvantages of mobile computing devices. The energy consumption of a mobile device depends on the power of electronic equipments installed on the mobile device, for example types of hard disks or CPU. Moreover, the battery life also depends on the number of applications and the application types that operate on the mobile

1. Sources: www.irda.org, www.bluetooth.org, www.ieee802.org, and www.intel.com
2. IrDA stands for Infrared Data Association
3. Yet to be standard devices. For example, a mobile phone can live up to five days but a laptop can only be able to operate for several hours; text processing applications consume less power than graphical applications. Transaction processes that are being carried out at a mobile host can be interrupted or re-scheduled if the mobile host is exhausting its power supply.

Limited Storage Capacity

The storage capacity of a mobile computer (i.e., hard disks or memory) is much less than a stationary computer and is harder to be expanded. Therefore, a mobile host may not be able to store the necessary data that is required for its operations in disconnected mode. Consequently, transaction processes on the mobile host will be delayed due to data unavailability, or require longer processing time due to frequent memory swapping operations.

Limited Functionality

The functionality of mobile devices is also limited in terms of the graphical user interface, the application functionalities, and the processing power.

Therefore, a mobile host may be unable to perform some of transaction operations, or requires longer processing time to perform these operations. For example, a small PDA may only be able to view black and white pictures.

Unreliable Equipment

The data stored at a mobile host can be lost if a catastrophic failure happens. This could heavily impact the durability property of transactions because of the loosing of the committed results of transactions that are stored at the mobile host. To avoid this problem, data stored at mobile hosts must be backed-up at stationary database servers as much and as soon as possible.

THE BEHAVIOR OF MOBILE HOSTS IN MOBILE ENVIRONMENTS

In mobile environments (Myers 1997), mobile transactions are initiated and/ or processed at mobile hosts. The mobile hosts can participate in the mobile transaction execution processes in different ways. First, a mobile host can initiate a mobile transaction, submits the transaction to appropriate (non-mobile or mobile) hosts for processing, and receives the committed results. In this way, the mobile host plays as four Sources as a terminal transaction client. Second, a mobile host can take part in the actual transaction execution process, i.e., the entire or part of a mobile transaction is carried out by the mobile host. The mobile host plays a vital role in the transaction execution process. Therefore, we need to answer the following question: How do the characteristics of the mobile environments affect the behavior of the mobile host? The behavior of mobile hosts in mobile environments is categorized into two dimensions:

First, the movement of the mobile host is affected by both the requirements of the mobile tasks and the environmental conditions. Second, the operation of the mobile host depends on its internally designed capacity and externally associative factors. For example, the performance of computational operations depends on the available energy of the mobile host's battery, and the network operations rely on both the connectivity capacity of the mobile host and the provided network services. The behavior of mobile hosts is discussed in the following.

Movement of Mobile Hosts

The movement behavior of a mobile host (Noble, Price, and Satyanarayanan 1995) describes the actual mobility states of the mobile host. While operating in mobile environments, the mobile host can be either in stopping or moving state. The two movement states are explained as follows:

Stopping

A mobile host is said to be in stopping state either when its movement velocity is zero, or when the location of the mobile host is not considered changing within a period of time. For example, as bus stops at a bus-stop to pick up passengers, a salesman is selling products at a shopping centre, or two mobile hosts are always moving close to each other.

Moving

A mobile host is in moving state either when its movement velocity has a value greater than zero, or when the location of the mobile host is considered changing over time. For example, a bus is moving along a road or a salesperson travels to several places during the day. While in moving state, the mobile host can continuously change its velocity and direction of movement.

On the one hand, the movement behavior of a mobile host can affect the mobile tasks that are carried out by the mobile host, e.g., a public transport vehicle needs to strictly follow a timetable. On the other hand, the movement of the mobile host can be affected by the surrounding environment conditions, e.g., traffic jam. The movement behavior of the mobile host demands additional supports such as location management, and awareness of location dependent data.

Operations of Mobile Hosts

The operation behavior of mobile hosts depends on the availability of mobile resources such as network connectivity and battery energy. We distinguish two operation modes for mobile hosts in mobile environments: isolation and interaction. These operation modes of the mobile hosts are explained as follows:

Interaction

When a mobile host is sharing data with other hosts, it is said to be in an interaction mode. The two essential prerequisite conditions for the interaction mode are:

1. The mobile host is operational
2. The network connectivity is available. It is not necessary that the mobile host always connects to other hosts all the times. This can help the mobile host to save the battery energy and to reduce communication cost. However, in an interaction mode, the communication channel between the mobile host and other hosts must always be available and establish-able whenever it is needed.

Isolation

When the communication channel between a mobile host and other hosts is not available, the mobile host is disconnected from other hosts and is said to be in an isolation mode. There are many factors that contribute to disconnection of the mobile host, for example the mobile host moves out of the wireless communication range, or network services are not available, or the mobile host is running out of its energy. The isolation mode can be further refined to autonomous and idle sub-modes.

Autonomous

When a mobile host operates by itself, it is said to be in autonomous mode. In the context of mobile transaction processing, we refer this mode as disconnected processing mode.

Idle. In this mode, the mobile host is not able to operate or has to delay its operations. The behavior of mobile hosts also illustrates the correlations among the three characteristics of the mobile environments. Disconnections in communication can be the results of the mobility of mobile hosts and/or the limitation of mobile resources. When mobile hosts communicate with others via short-range wireless network technologies, e.g., infra-red or Bluetooth or wireless LAN, the communication will be disconnected if the mobile hosts move outside the limited communication range. The mobile hosts can be disconnected for short periods, i.e., seconds or minutes, and more frequently when they are moving in and out of the shadow of physical obstructions such as high buildings. The disconnection period can be long, i.e., hours or days, when the mobile hosts stay in some locations in which the wireless network service is not available. The mobile hosts can also volunteer to disconnect if the supplied energy is running out. On the other hand, the heavy use of network activities can shorten the battery life of the mobile host.

TRANSACTION PROCESSING IN MOBILE ENVIRONMENTS

The main differences between the mobile environments and distributed environments are:

1.   Mobile computing hosts
2.   Wireless networks.

The mobile hosts usually have less computing resources and capacity than stationary hosts. For example, a laptop computer has lower processing speed and smaller storage capacity than a desktop computer, and its operation might depend on the limited battery energy. Consequently, it takes longer time for a transaction to be processed at a mobile host (Noble, Satyanarayanan, Narayanan, Tilton, Flinn, and Walker 1997). Moreover, mobile computers are easily damaged, i.e., less reliable. The results of committed transactions, which are stored at a mobile computer, can be lost if the mobile computer is damaged, i.e., the durability property of transactions may not be fully guaranteed. Therefore, the committed results of transactions in mobile environments should additionally be saved at the stationary hosts as in distributed environments. The movement of mobile hosts brings additional requirements and demands that the mobile transaction processing system must handle, for example hand-over processes or locally dependent data. In distributed environments, these demands do not exist. Mobile computing hosts communicate with other hosts via wireless networks. Compared to a wired network, a wireless network is usually less reliable, i.e., disconnections can occur frequently; has lower bandwidth, i.e., megabits versus gigabits; and is limited in communication range, i.e., mobile hosts must stay within limited distance to be connected. Because of these unique features of wireless networks, it can take longer time to download necessary data into the local storage devices at the mobile hosts; or due to disconnections, the mobile hosts will not be able to obtain the needed data. Consequently, transactions in mobile environments may experience long blocking periods and inconsistent data.

In mobile environments, transaction processing systems consist of both mobile and non-mobile hosts, and can be divided into two different layers. The distributed transaction processing layer corresponds to the execution of mobile transactions that are carried out on non-mobile hosts. The mobile transaction processing layer corresponds to the execution of mobile transactions that are carried out on a mobile host or distributed among mobile hosts. Due to the above distinguishing and challenging characteristics of mobile environments, transaction processing in mobile environments is more difficult than in distributed environments, especially in terms of concurrency control, data availability, and recovery mechanisms.

ARCHITECTURE OF MOBILE TRANSACTION ENVIRONMENTS

In this section, we discuss the architecture of the mobile transaction environments. In general, the mobile transaction environments include three different components: mobile hosts (MH), mobile support stations (MSS) and fixed hosts where database servers (DB) reside. A mobile environment is a geographical territory. The geographical territory is divided into a collection of areas called mobile cells. Wireless communications in each mobile cell is provided by a single low-power transmitter-receiver. There might be some areas in the mobile environments in which the wireless communication is not available. This could be caused by the limited service of the wireless communication providers or the structural of physical objects in the areas, for example concrete tunnels or remote islands. The geographical mobile environment, therefore, can be considered as a collection of mobile cells that are separated or overlapped with others. The size of mobile cells is not necessarily equal, due to the differences of operational power of the transmitter-receiver devices. The wireless technologies that are provided in each mobile cell can be different, for example wireless LAN or wireless USB. As a consequence, network bandwidth, network latency, communication protocols and covered ranges are different among mobile cells.

In each mobile cell, there is a special computing host called the mobile support station. The role of the mobile support station is to provide additional computing services to all the mobile hosts that currently reside in the mobile cell.

Mobile hosts are portable mobile computing devices which have the capability to cache a limited amount of information. Database servers are stationary computers that are connected via high speed wired-networks, and play roles as permanent data storage repositories. Shared data is distributed on these database servers. Mobile support stations (also called base stations) are stationary or mobile computers. Mobile support stations have higher processing power and data storage capacity than the mobile hosts. The role of the mobile support stations is to support mobile hosts communicating with other mobile hosts or database servers. Mobile hosts communicate with the mobile support stations via the wireless networks. Communications between the database servers and the mobile support stations are via wired networks or dedicated wireless connections. Mobile hosts move in mobile environments while carry out mobile tasks. While being in a mobile cell, a mobile host can be either connected or disconnected with the mobile support station of this mobile cell. The mobile host may only connect to the mobile support station when there is a need for sharing of data. This will help to save the limited energy of the mobile host and to reduce the communication cost. On the other hand, because of the limitations of wireless networks, a mobile host may not always be able to establish a communication channel with the mobile support station. If a mobile host is in the area that is an intersection of two or more mobile cells, it can connect to any mobile support station.

The mobile hosts can move within one mobile cell or across a large area covered by several mobile cells. When a mobile host is leaving a mobile cell and entering a new mobile cell, the communication channel and other related information between the mobile host and the previous mobile support station will be transferred to the next mobile support station. This process is called hand-over or hand-off process. The new mobile support station at the new mobile cell will continue carrying out the support to the mobile host.

However, it is not necessary that hand-over processes must happen every time the mobile host enters a new mobile cell. For example, the mobile host can operate in an autonomous mode when the wireless network is not supported in the new mobile cell. Furthermore, a mobile host does not have to disconnect from the old mobile support station before it can connect to the new mobile support station. A mobile host can connect to a new mobile support station while connecting to the old mobile support station. The hand-off process can be planned beforehand if the travel route of the mobile host is known in advanced and strictly followed. Otherwise, the hand-off process can only be carried out after the mobile host has established a connection with the new mobile support station, i.e., after the new destination of the mobile host is known.

Mobile cells one and two are separated, while mobile cells three and four are overlapped. A mobile host moves from position A in mobile cell one to position B in mobile cell four. The travel route of the mobile host passes through mobile cells two and three. When the mobile host is leaving cell one, it will enter a disconnected interval in the area between the mobile cells one and two. While in the mobile cell two, the mobile host will be supported by the mobile support station that is a dedicated mobile host. When the mobile host is in the mobile cell three, it may not connect to the mobile support station all the time. In the intersection region of the mobile cells three and four, the mobile host can connect to the mobile support station of either mobile cell three or mobile cell four. The hand-over processes occur when the mobile host moves from one mobile cell to another along the travel route.

Characteristics of Mobile Transactions

Transactions in mobile environments (Object Management Group, Inc. CORBA Services 1997) possess many challenging characteristics due to the characteristics of the mobile environments. In this section, we will discuss the characteristics of mobile transactions. The characteristics of mobile transactions are described as follows:

Mobility of Transactions

The execution of transactions in mobile environments is tightly coupled with the behavior of the mobile hosts. A mobile host can initiate mobile transactions or participate in the transaction execution processes. When a mobile host moves from one location to another, all the transactions that are being carried out at that mobile host will also move. Consequently, many computing activities associated with these transactions are moved or changed, for example handling hand-over processes, establishing new communication channels, or updating the routing tables. In other words, the mobility of transactions causes the movement of related transaction resources, controls, and services.

Long-Lived Transactions

Transactions in mobile environments have longer life (i.e., long-lived) than traditional ACID transactions. This is due to the overheads that are caused by two aspects: the data availability and the execution interruptions.

Data Availability

In mobile environments, the data availability at a mobile host can be affected by many factors. First, the movement of the mobile host causes the movement of related information. This will cause additional overhead to the transaction execution time. Second, the bandwidth of wireless networks is limited; therefore it will take longer time to obtain the necessary data. Third, the mobile computing devices have limitations in storage capacity; therefore, the mobile host may not able to cache the required information to support disconnected transaction processing. In addition, due to the unexpected disconnections of the wireless networks, a transaction will not be able to release the controls on shared data to Transactions at other hosts as scheduled; this means that this transaction blocks the execution of other transactions.

Execution Interruptions

The execution of transactions can be interrupted while being carried out at the mobile host. The interruptions can be caused by either the surrounding environment conditions or the limitation of computing capacity of the mobile host. For example, a wireless network disconnection suddenly occurs during the execution of transactions, or the performance of the mobile host is slowing down due to heavy computing load. The interruptions can happen frequently and cause transactions to be suspended or aborted.

Adaptive Transaction Processing

Due to the real-time movement of the mobile hosts, the limitations of the wireless networks, and the variation of the mobile resources, the execution plan of a transaction in mobile environments may not be as scheduled. Therefore, the mobile transaction processing system must have the ability to support adaptive transaction processing that includes: distributed and disconnected transaction processing.

Distributed Transaction Processing

Due to the limitations of processing capacity and resources, mobile hosts require additional support from other hosts to carry out transactions. For example, a transaction, which is initiated by a mobile host and accesses a large data set that is not cached at the mobile host, could be moved to stationary hosts for executing. This could reduce transaction processing time and avoid transferring a large amount of data through a slow wireless network, i.e., achieving higher throughput for the transaction processing system. Furthermore, the portable computing devices are easily damaged; therefore, the results of committed transactions must be saved at stationary database servers.

Disconnected Transaction Processing

A mobile host can be disconnected from the database servers for long periods; therefore, transactions that are executed at the mobile host may suffer from long blocking if the necessary data is not available at the mobile host. To deal with this problem, the mobile transaction processing system should have the capacity to cache enough data so that it can carry out the transactions while being disconnected from the database servers.

Temporary Data Inconsistency

Due to long disconnection periods, shared data among different mobile hosts may not be fully consistent all the time. For example, a transaction at a disconnected mobile host can modify a shared data item that is currently being read-only cached in a local storage of another disconnected mobile host. Data synchronization processes will be carried out when the disconnected mobile hosts reconnect to the database systems so that the data consistency of the database systems will be achieved.

Heterogeneous Processing

Many types of mobile devices can be involved in transaction execution processes. Interactions or communications among participating parties are carried out via the support of different types of wireless network technologies and protocols. Furthermore, different database systems are accessed during the execution of mobile transactions. All these factors contribute to the heterogeneous processing characteristic of mobile transactions.

TRANSACTION MODELS

In this chapter, we survey several selected transaction models and transaction processing systems that have been purposely developed to support transaction processing in mobile environments. We will also recap some traditional transaction models whose features could be used in the mobile environments.

Traditional Transaction Models

As the transaction environment (Ott, Michelitsch, Reininger, and Welling 1998) evolves from the centralized environment to distributed and mobile environments, the properties and the structure of transactions change. However, several basic transaction models are indispensable. In other words, they are still useful and applicable in the new mobile environments. In this section, we will review the following transaction models:

1.   Flat transaction model
2.   Nested transaction model
3.   Multilevel transaction model
4.   Sagas transaction model
5.   Split and Join transaction model

For each transaction model, we briefly describe the transaction model, the properties and discuss how the features of the transaction model could be used in the mobile environments.

Flat Transaction Model

The flat transaction model presents the simplest transaction structure that fully meets the ACID properties. The building block of a flat transaction, between Begin and Commit /Abort operations, contains all the database operations that are tightly coupled together as one atomic database operation. The flat transaction begins at one consistent database state, and either ends in another consistent state, i.e., the transaction commits, or remains in the same consistent state, i.e., the transaction aborts. The flat transaction model fully meets the standard ACID properties. The flat transaction is fully isolated during its execution, and any failure causes the whole transaction to abort. The results of a committed flat transaction are durable and permanent. Due to the strict ACID properties, the flat transaction model is not suitable in mobile environments. However, the flat transaction model plays an important role for building more advanced transaction models. For example, a complicated transaction model can consist of a set of smaller flat transactions. The flat transaction model can be easily supported at the application programming level.

Nested Transaction Model

The nested transaction model defines the concepts and the mechanisms for breaking up the large building block of a flat transaction into a set of smaller transactions, called sub-transactions. Thus, the nested transaction model has a hierarchical tree structure that includes a top-level transaction and a set of sub transactions (either parent or children transactions). Sub-transactions at the leaf level of the transaction tree are flat transactions.

The nested transaction model has the following characteristics. First, children transactions are flat transactions. Second, the children transactions start after their parent have started, and can autonomously commit or abort. However, the results of the committed children transactions do not take effect until their parent transactions commit. In other words, the nested transaction only commits when the top level transaction commits. And third, when a child transaction commits, its results become visible to its parent transaction. If a parent or the top-level transaction aborts, all the sub-transactions must abort, regardless of their states. The concept of the nested transaction model can be applied in mobile environments, especially for decomposing a large transaction into sub transactions which can be carried out concurrently.

Multilevel Transaction Model

The multilevel transaction model is looser than the nested transaction model in terms of the relationship between parent and children transactions. Sub-transactions in the multilevel transaction can commit or abort independently of their parents. This is supported by the concepts of compensating transactions. We will briefly discuss the concept of compensating transactions, and its opposed contingency transactions. Compensating transactions are designed to undo the effect of the original transactions that have aborted. The compensating transactions are triggered and started when the original transactions fail. Otherwise, the compensating transactions are not initiated. Once a compensating transaction has started, it must commit. In other words, the compensating transactions cannot abort. If a compensating transaction fails, it will be restarted. Contingency transactions are designed to replace the task of the original transactions that have failed. Contingency transactions are also triggered by the failures of the original transactions. Note that it is not always possible to specify the compensating or contingency transactions for an original transaction.

The isolation property is relaxed in multilevel transaction model. The committed results of sub-transactions are visible to other transactions. The atomicity property is ensured by the means of compensating transactions. The multilevel transaction model is applicable in mobile environments. Multilevel transaction model not only relaxes the isolation property of transactions but also provides a flexible recovery mechanism by the means of the compensating and contingency transactions.

Sagas Transaction Model

The Sagas transaction model also makes use of the concept of compensating transactions to support transactions whose execution time is long. A Sagas transaction consists of a consecutive chain of flat transactions S_i that can commit independently. For each flat transaction S_i, there is a compensating transaction CP_i that will undo the effect of the transaction S_i if the transaction S_i aborts. A compensating transaction CP_i in the Sagas chain is triggered by the associated transaction S_i or the compensating transaction CP_{i+1}. If the Sagas transaction commits, no compensating transaction CP_i is initiated, otherwise the chain of compensating transactions is triggered.

The unit of control of a Sagas transaction is the whole transaction chain. Sagas relaxes the isolation property by allowing component transactions Si to commit. The atomicity property of Sagas is achieved by the commitment of the last transaction component Sn in the chain or by the backward execution of the compensating transaction chain. The Sagas transaction model is useful in mobile environments because of its ability for supporting transactions which are long-lived. The isolation property is also compromised. Therefore, the concept can be used to support sharing of data during the execution of mobile transactions. Moreover, it is possible to modify the Sagas model so that we can minimize the closing of useful work when a component transaction Si aborts, for example by deploying contingency transactions instead of compensating transactions. The main drawback of Sagas is the sequential execution of component transactions in the chain.

Split and Join Transaction Model

The Split and Join transaction model was proposed to support the open ended activities that associate with transactions. The Split and Join transaction model focuses on activities that have uncertain duration, uncertain developments, and are interactive with other concurrent activities. The main idea is to divide an on-going transaction into two or more serializable transactions, and to merge the results of several transactions together as one atomic unit. In other words, the Split and Join transaction model supports reorganizing the structure of transactions.

The Split and Join transaction model divides the accessed data set of a transaction into different subsets that will be used by newly created and serializable transactions. The goal is to commit part of the original transaction and to make committed results or resources available to other transactions. The Split and Join transaction model benefits transactions in mobile environments in terms of dynamic re-structuring of transactions.

Mobile Transaction Models

We have reviewed several traditional transaction models whose features are still useful in mobile environments. The traditional transaction models, however, do not have the ability to deal with other challenging requirements of mobile transactions, such as supporting the mobility of transactions and coping with disconnections. Consequently, there are many advanced transaction models that have been developed to particularly support mobile transactions. In this section, we will review several selected mobile transaction models that have the ability to efficiently support mobile transactions. The follows mobile transaction models will be surveyed:

1. Report and Co-transaction model
2. Pro-motion transaction model
3. Two-tier transaction model
4. Weak-Strict transactions model
5. Pre-write transaction model
6. Pre-serialization transaction model
7. Kangaroo transaction model
8. Moflex transaction model
9. Adaptable mobile transaction model (MTS)

For each model, we describe the transaction model and its properties, then we address how the model:

1. Handles the mobility of transactions
2. Deals with disconnections
3. Supports distributed transaction execution among mobile and non-mobile hosts.

Reporting and Co-Transaction Model

Reporting and Co-transactions transaction model is based on a two level nested transaction model. A reporting transaction TR shares its partial results to top-level transaction S by delegating its operations. The delegation process can happen at any time during the execution of transaction TR. A co-transaction is a reporting transaction but it cannot continue executing during the delegation process. Thus, the co-transaction behaves as a co-routine, and resumes execution when the delegation process is completed. The top-level transaction is the unit of control, and atomic sub transactions are compensable transactions. A Reporting transaction that is compostable does not have to delegate all of the committed results to the top-level transaction when it commits. Sub-transactions that are non-compensable delegate all of their operations to the top-level transaction when it commits.

The locations of mobile hosts are determined via the identification of mobile support stations. However, the model does not mention explicitly what happens when mobile hosts move from one mobile cell to another. Delegation operations require a tight connectivity between the delegator (i.e., Report and Co-transaction) transactions and the delegate transaction; therefore, disconnection is not supported in this model. The model supports distributed transaction processing among mobile hosts and fixed hosts where the network connectivity among these hosts is assumed to be available when it is needed.

Pro-Motion Transaction Model

The Pro-motion transaction model is a nested transaction model. The Pro-motion model focuses on supporting disconnected transaction processing based on the client-server architecture. Mobile transactions are considered as long and nested transactions where the top-level transaction is executed at fixed hosts, and sub transactions are executed at mobile hosts. The execution of sub-transactions at mobile hosts is supported by the concept of compact objects.

Compact objects are constructed by compact manager at database servers. Necessary information is encapsulated within a compact object. The compact objects are co managed by the compact managers (resided at the database servers), the mobility managers (at the mobile support stations), and the compact agents (at the mobile hosts). The compact object plays a role as a contractor that supports data replication and consistency between mobile hosts and database servers. When a mobile host is disconnected, the compact agent takes responsibility for managing all local database operations of mobile transactions at the mobile host. When the mobile host reconnects to database servers, the compact objects are verified against global consistency rules before the locally committed mobile transactions are allowed to commit. Transaction processing consists of four phases: hoarding, disconnected, connected, and resynchronization. Shared data is downloaded to the mobile host in the hoarding phase. When the mobile host is disconnected from the fixed host, transactions are disconnectedly executed at the mobile host. If the mobile host connects to the fixed database, the transactions are carried out with the support of the compact manager. When the mobile host reconnects to a fixed host, the results of local transactions are synchronized with the database.

The Pro-motion transaction model supports ten different levels of isolation. Transactions are allowed to locally commit at mobile hosts; the committed results of these transactions are made available to other local transactions. However, the local committed results must be validated when the mobile hosts reconnect to the database servers. Therefore, the durability property of transaction is only ensured when the transaction results are finally reconciled at the fixed database. Though the mobility manager supports communications between the mobile host and the database servers, how the Pro-motion transaction model supports transaction mobility is not explicitly discussed. Pro-motion transaction model supports disconnected transaction processing via the support of compact objects. When the mobile host is disconnected from the fixed database, the sub-transactions are split and executed at the mobile host (these split sub-transactions are not joined when the mobile host reconnects to the fixed database). Disconnected transaction processing is a dominant transaction processing mode in Pro-motion even when the mobile hosts are able to connect to the database server. Therefore, the Pro-motion transaction model requires high-capacity mobile resources at the mobile hosts. Transactions are mostly executed at mobile hosts and the results are reconciled at the database servers. Therefore, the distributed transaction processing is not strongly supported by the model.

Two-Tier Transaction Model

The two-tier (also called Base-Tentative) transaction model is based on a data replication scheme. For each data object, there is a master copy and several replicated copies. There are two types of transaction: Base and Tentative. Base transactions operate on the master copy; while tentative transactions access the replicated copy version. A mobile host can cache either the master or the copy versions of data objects. While the mobile host is disconnected, tentative transactions update replicated versions. When the mobile host reconnects to the database servers, tentative transactions are converted to base transactions that are re-executed on the master copy. If a base transaction does not fulfill an acceptable correctness criterion (which is specified by the application), the associated tentative transaction is aborted.

Tentative transactions locally commit at the mobile host on replicated copies, and the committed results are made visible to other tentative transactions at that mobile host. The final commitments of those tentative transactions are performed at the database servers. Two-tier transaction model does not support the mobility of transactions. While the mobile hosts are disconnected from the database servers, tentative transactions are locally carried out based on replicated versions of data objects. Two distinct transaction execution modes are supported: connected and disconnected. Transactions are tentatively carried out at disconnected mobile hosts, and re-executed as base transactions at the database servers.

Weak-Strict Transaction Model

The Weak-Strict (also called Clustering) transaction model consists of two types of transaction: weak (or loose) and strict. These transactions are carried out within the clusters that are the collection of connected hosts which are connected via high-speed and reliable networks. In each cluster, data that is semantically related is locally replicated. There are two types of a replicated copy: local consistency (weak) and global consistency (strict). The weak copy is used when mobile hosts are disconnected or connected via a slow and unreliable network. Weak and Strict transactions access weak and strict data copies, respectively. When mobile hosts reconnect to database servers, a synchronization process reconciles the changes of the local data version with the global data version. Weak transactions are allowed to commit within its cluster, and results are made available to other local weak transactions. When mobile hosts are reconnected, the results of weak transactions are reconciled with the results of strict transactions. If the results of a weak transaction do not conflict with the updates of strict transactions, weak transactions are globally committed; otherwise they are aborted.

The concept of transaction migration is proposed to support the mobility of transactions, and to reduce the communication cost. When the mobile host moves and connects to a new mobile support station, parts of the transaction that are executed at the old mobile support stations are moved to the new one. However, no further details about the design or implementation are given.

The Weak-Strict transaction model supports transaction processing in disconnected and weakly connected modes via weak transactions. Transaction execution processes can be distributed between the mobile host and the database servers within a cluster that the mobile host participates in. However, the distributed transaction processing among mobile hosts in a cluster is not discussed.

Pre-Write Transaction Model

The Pre-write transaction model was proposed to increase data availability in mobile environments. Mobile transactions are transactions that are initiated at the mobile host. Pre-write transaction model aims to increase the data availability at mobile hosts. This is achieved by allowing a transaction on a mobile host to submit pre-write operations that write the updated data values, and then issue a pre commit state to the mobile support station. After that, the rest of the mobile transaction can be carried out and finally committed at fixed hosts. The small variation, which is specified by the applications, between the pre-committed result and the final committed result is acceptable. Pre-committed data values are accessible to other transactions via pre-read operations. Two different types of lock, which are the pre-read and pre-write, are introduced to support the new operations. Mobile transactions are not allowed to abort after they have submitted pre-commit operations to the mobile support station. This mobile transaction model can be used to support mobile hosts which have little or no capacity for transaction processing.

After a mobile transaction submits a pre-commit request, the pre-write values of the mobile transaction are made available to transactions. And the pre-committed mobile transaction is not aborted in any case. The final commitments of mobile transactions will be carried out by fixed hosts. The final committed and the pre-committed data values may not be identical. The roles of the mobile support station are to accept and to process pre-write and pre-commit operations submitted from the mobile host. When moving into a new mobile cell, a mobile transaction connects to the mobile support station in order to submit its pre-write and pre-commit operations.

Disconnected transaction processing is supported in the Pre-write transaction model. The mobile transaction is executed at the mobile host until the pre-commit state is reached. The major part of the mobile transaction is migrated to the fixed hosts via the mobile support station to be executed there. The mobile host partly takes part in the execution process until the pre-commit states of the mobile transaction are achieved. After this, the mobile host plays no role in the execution of the mobile transaction.

Pre-Serialization Transaction Model

Pre-serialization transaction model is built on top of local database systems. Mobile transactions (also called global transactions) are submitted from mobile hosts through the global transaction coordinators that reside at the mobile support stations. The mobile transaction is entirely processed at local database systems. At each node (or site), there is a site manager that administrates all the transactions executed at that node. When a global transaction is prepared to commit, a global transaction coordinator will carry out an algorithm, called Partial Global Serialization Graph algorithm that detects any non-serializable schedule among the mobile transactions. If there is a cycle in the graph, i.e., the schedule is non-serializable, the mobile transaction is aborted.

Each sub-transaction of a global transaction is managed by the local transaction manager. The global serializable graph of transactions is constructed by collecting sub-graphs from the local sites. The atomicity property of the global transaction is relaxed by the concepts of vital and non-vital sub-transactions. If a vital sub-transaction aborts, its parent transaction must abort. However, the parent transaction does not abort if a non-vital sub-transaction aborts. When a sub-transaction commits at the local database system, the results are made visible to other transactions at this local database system.

The global transaction coordinators that reside at the mobile support stations support the mobility of mobile transactions. This is done by transferring the global data structure from one global transaction coordinator to another as the mobile host moves from one mobile cell to another. Mobile transactions are submitted from a mobile host, and sub transactions are executed at local database servers. When the mobile host is disconnected, the global transaction is marked as disconnected if the disconnection is known and planned. The execution of the global transaction is still carried out at the local database servers. On the other hand, if the disconnection is unplanned, the global transaction is suspended. The global transaction is resumed when the mobile host reconnects to the mobile support station. Mobile transactions are submitted from mobile hosts, and the entire transactions are distributed among local database servers through the support of mobile support stations. The mobile hosts do not take part in the execution processes.

Kangaroo Transaction Model

The Kangaroo transaction model is designed to capture the movement behavior and the data behavior of transactions when a mobile host moves from one mobile cell to another. This transaction model is built based on the concepts of global and split transactions in a heterogeneous and multi-database environment. The global transaction is split when the mobile host moves from one mobile cell to another, and the split transactions are not joined back to the global transaction. The Kangaroo transaction model assumes that the mobile transactions may start and end at different locations. The characteristics of the Kangaroo transaction model are

1.   Mobile transactions that include a set of sub-transactions called global and local transactions are initiated by mobile hosts. These mobile transactions are entirely executed at the local database servers that reside on the fixed and wired connected networks.
2.   The execution of a Kangaroo sub-transaction in each mobile cell is supported by a Joey transaction that operates in the scope of the mobile support station. The Joey transaction plays role of a proxy transaction to support the execution of the sub transactions of the Kangaroo transaction in the mobile cell.
3.   The movement of the mobile host from one mobile cell to another is captured by the splitting of the on-going Joey transaction at the old mobile support station and the creating of new Joey transaction at the new mobile support station. The execution of the Joey transaction is supported by the Data Access Agents (DAA) that act as the mobile transaction managers at the mobile support stations.

The Kangaroo transaction is the basic unit of computation in mobile environments. The serializability of mobile transactions is not guaranteed, and there is no dependency among Joey transactions, i.e., each Joey transaction can commit independently. Two transaction processing modes, which are compensating and split modes, are supported by the model. For compensating mode, when a failure occurs, the entire Kangaroo transaction is undone by executing compensating transactions for all those Joey transactions. For split mode, the local DBMS takes responsibility for aborting or committing sub-transactions. The Kangaroo transaction model keeps track of the movement of mobile hosts via the support of the DAA that operates at the mobile support station. In other words, the mobility of mobile hosts is captured on the condition that the mobile hosts always may communicate with the mobile support stations. While mobile hosts move from one mobile cell to another, the hand-off processes are carried out by the DAAs.

Disconnected transaction processing is not considered in Kangaroo transaction model. The processing of Kangaroo transactions is entirely moved to the fixed database servers for executing. The mobile transactions are initiated at the mobile hosts, and entirely executed at fixed hosts. Transaction results are forwarded back to the mobile hosts. The Kangaroo transaction model has shown that the structure of mobile transactions at the specification and execution phases (with the dynamic support of Joey transactions) can be different because of the mobility behavior, i.e., fast or slow movements, of the mobile host.

Moflex Transaction Model

The Moflex transaction model is an extension of the Flex transaction model to support mobile transactions. The Moflex model is built on top of multi-database systems and based on the concepts of split-join transactions. The main characteristics of a Moflex transaction are:

1. A Moflex transaction that consists of compensable or non-compensable sub transactions is initiated by the mobile host. These sub-transactions are submitted to the mobile transaction manager (MTM) that resides at the mobile support station. The MTM will send these sub-transactions to the local execution monitor (LEM) at local database systems for executing.

2. Each Moflex transaction T is accompanied by a set of success and failure transaction dependency rules, hand-over control rules and acceptable goal states. Dependent factors that include the execution time, cost and execution location of transactions are also specified in the definition of the Moflex transaction. Furthermore, joining rules are provided to support the join of the split sub transactions (sub-transactions are split when the mobile host moves from one mobile cell to another).

The mobile transaction managers make use of the two-phase commit protocol to coordinate the commitment of the Moflex transaction. The Moflex transaction commits when its sub-transactions that are managed by MTM have reached one of the acceptable goal states, otherwise it is aborted. A compensable sub-transaction is locally committed, and the results are made visible to other transactions. For no compensable sub-transactions, the last mobile transaction manger, which corresponds to the end location of the mobile host, plays the role as the committing coordinator. The mobility of transactions is handled by splitting the sub-transaction, which is executed on the local database at the current mobile cell, as the mobile host moves from one mobile support station to another (with the support of the mobile transaction manager). Hand-over control rules must be specified for each sub-transaction. If a sub-transaction is compensable and location independent, it will be split into two transactions; one will continue and commit at the current local database, the second will be resumed at the new location. If the sub-transaction is location dependent, at the new location, the sub-transaction must be restarted. If a sub-transaction is non-compensable, the sub-transaction is either restarted as a new one in the mobile cell if it is location dependent, or continued if it does not depend on the location of the mobile host.

Adaptable Mobile Transaction Model

An adaptable mobile transaction model and a mobile transaction service (MTS) are proposed to support the adaptability of mobile transaction execution. The MTS architecture is a three-tier client/agent/server one in which the clients are mobile hosts, the agents reside at mobile support stations, and the servers are fixed database servers. The main goal of the MTS is to adapt the transaction execution to different environment conditions. The adaptive mobile transaction consists of component transactions T_i, compensating transactions CT_i and the execution strategy ES. The execution strategy is a list of execution alternatives comprised of environment descriptors ED and component transactions. Changes of environment conditions are captured via an event notification service.

Only one execution alternative of the adaptive mobile transaction is executed at any moment. The component transactions are ACID transactions which can belong to one or more execution alternatives. Changes in the execution alternatives may result in the abortion of the component transactions. If a component transaction belongs to different execution alternatives, the component transaction is continued with the new execution alternative.

The mobility of transactions is not defined by the adaptable mobile transaction. However, the hand-off process is treated as a change of environment conditions via an e-hand-off event. The disconnected processing of mobile transactions is specified in execution alternatives, and is applied when an e-disconnection event occurs. The mobile transaction service defines five different execution modes that specify how a mobile transaction could be executed among the fixed database servers and the mobile hosts. The adaptable mobile transaction takes into account dynamic changes of mobile environments, and supports different execution alternatives in accordance with the environment conditions. The main disadvantage of the model is that the execution alternatives must be specified in advance.

Modes Distributed Execution

1. Entirely at database servers
2. Entirely at the mobile host
3. At one mobile host and several DB
4. At several mobile hosts
5. At several mobile hosts and DBs

ISSUES RELATED TO MOBILE TRANSACTION PROCESSING SYSTEMS

We discuss issues that are related to mobile transaction processing systems. The three issues are: mobile database replication, advanced transaction commitment protocols, and mobile data sharing mechanisms. These three issues contribute in a vital way to the performance of transaction processes in mobile environments.

Mobile Database Replication

In mobile environments, to cope with the disconnections of the wireless networks, the mobile hosts must be able to cache necessary data to support disconnected transaction processing. A database portion that is cached at a mobile host is called the mobile database. Mobile databases offer higher level of data availability at disconnected mobile hosts; thus, enhance the performance of mobile transaction processing systems. Before the mobile hosts are disconnected from the database servers, shared data is cached at the mobile host. When the mobile host is disconnected from the database servers, cached data is modified. When the mobile hosts reconnect to the database server, shared data that has been modified at the local cache will be reconciled with the original versions.

Keeping data consistency among these copies all the time is difficult. Therefore, the main issue is how to avoid or be aware of data inconsistency among such copies. This can be achieved by several ways, for example an advanced locking protocol or signoff/ check-in/check-out operations. The pre-write lock is an additional lock layer that is deployed at the mobile support station to support mobile transactions to access shared data, i.e., without connecting to the database server. Transactions can connect to either the database server or the mobile support station to access shared data. If a shared data is modified, it will first be stored at the mobile support station before being updated in the database server. For sign-off/check-in/check-out operations, consistent shared data is downloaded from the database server to mobile hosts via the support of a proxy-transaction (called a pseudo-transaction) to support disconnected transaction processing. When the mobile hosts reconnect to the database server, mobile transactions will be checked to ensure that they are serializable with other transactions at the database server.

The mobile databases must be able to support mobile hosts to cope with different types of disconnections. There are two forms of disconnection: planned and unplanned. For planned disconnections, the mobile hosts inform the database servers about the disconnections so that the mobile databases can be well prepared. The strict mobile database replication model uses the standard shared and exclusive lock modes for controlling conflicting database operations among replicated copies. Relaxed mobile database replication model allows transactions to concurrently access replicated database portions at different mobile hosts as long as there is an acceptable execution schedule among involved transactions. For example, check-out with mobile read, checkout with system read, or relaxed check-out modes. To our knowledge, there is no mobile database model that supports mobile databases to deal with unplanned disconnections (which will be dealt with in our mobile transaction processing system).

Advanced Transaction Commitment Protocols

The standard 2PC protocol may not be appropriate in mobile environments because it is a possibly blocking protocol and requires many messages. There are more advanced transaction commitment protocols that have been proposed. The transaction coordinator that resides at the mobile support station will decide to commit or abort transactions based on a timeout value. The timeout value is the total of execution timeout and shipping timeout. A mobile transaction will be allowed to commit if all of the updates of sub-transactions are received by the coordinator before the timeout value is expired; otherwise the transaction is aborted. The timeout commit protocol requires that mobile hosts always connect to the mobile support station and the database servers. The main drawback of this protocol is that it is hard to define or estimate the execution and shipping timeout values in mobile environments.

The Unilateral Commit Protocol is proposed to support transaction commitment in disconnected mode. This protocol reduces the number of exchanged messages by removing the second voting phase of the standard 2PC protocol (thus, this protocol is also called a one-phase commit protocol). If a mobile transaction reaches the prepare-to commit phase, it will commit. There are other commit protocols that are developed by taking into account the special characteristics of mobile transactions. Some examples are the commitment of read-only transactions that are carried out separately from updating transactions (by exploring the special consistency requirements of read-only transactions), and the pre-commit protocol (by tolerating the difference between pre-committed and committed results which is specified by applications).

Mobile Data Sharing Mechanisms

In this section, we address the mechanisms that support sharing of data in mobile environments. In general, shared data is stored at dedicated non-mobile database servers. Mobile hosts need to connect to these database servers to access shared data. However, due to the disconnections in communication, the mobile hosts may not always be able to connect to the database servers. This leads to the demand of a temporary sharing workspace that is stored at dedicated hosts. Existing models that have been designed to support sharing data in distributed environments, for example the common-local workspaces model or the sharing tuple space, will not be suitable for mobile environments due to the static configuration and the lack of mobile and dynamic workgroup supports.

Recently, there are many research proposals that focus on supporting data sharing among mobile hosts in mobile environments. The two essential components that contribute to the mobile data sharing are:

1. The dynamic mobile workgroup management
2. The data access mechanisms

The dynamic mobile workgroup management focuses on the organization and management of temporary mobile workgroups that are the collection of mobile hosts. The data access mechanisms are based on either the client-server or the peer-to-peer architecture. The Accessing Mobile Database (AMDB) architecture is based on the concepts of mobile agents and the client-server model. The main idea is to form a Mobile Database Community (MDBC) in which mobile clients access mobile databases that are stored at dedicated mobile hosts. The LIME (Linda in Mobile Environments) architecture makes use of mobile agent technology to support sharing of data among different mobile hosts.

Survey of Commercial Products

We have reviewed several mobile transaction models that are mostly used for academic research purposes. There is little information about how these mobile transaction models are deployed in real application products. In this section, we review mobile transaction processing in commercial products. The following products are surveyed: Microsoft SQL Server CE, Oracle Lite [Ora], and IBM DB2 Everyplace. We focus on describing in detail how these commercial products support mobile transactions and how data consistency is achieved.

Microsoft SQL Server CE

The Microsoft SQL Server CE (SSCE) is a client-agent-server architecture that supports database applications on mobile hosts. The database on a mobile host is a small replicated portion of the main database. When mobile hosts are disconnected, transactions are processed locally at the mobile hosts. When mobile hosts reconnect to the database server, synchronization processes are carried out to reconcile information. The client agent at the mobile host connects to the server agent through the Internet Information Server (IIS) that resides on the database server. This means that the role of mobile support stations is not an issue in SSCE systems.

The Microsoft SQL Server CE supports both flat and nested transactions at mobile hosts. Sub-transactions only reveal committed results to the parent transaction. When the top-level transaction commits, the results are visible to local transactions at the mobile host. Transactions at mobile hosts are executed sequentially. When the mobile host reconnects to the database server, a synchronization process is performed to reconcile information. The client agent sends all changes in the local database to the server agent. The server agent, then, writes the updates to a new input file and initiates a reconciliation process at the SQL Server Reconciler. The reconciliation process will detect and resolve conflicts. Different conflict resolutions are supported in the SSCE system, for example priority based or user defined. When the reconciliation process completes, it will inform the SQL Server Replication Provider to finally write the successful updates to the database server. When there are updates at the database server, an inverse process is carried out to propagate these updates to the mobile host.

Oracle Lite

Oracle Lite is a client-server architecture that makes use of a replicated copy of the main database (which is called a snapshot) to support disconnected transaction processing at mobile hosts. Oracle Lite does not include mobile support stations in its architecture. The replicated database system at the mobile host is called a snapshot that can be read-only or updatable. When the mobile host is disconnected from the database server, transactions are processed locally. The snapshot is synchronized with the master copy at the database server when the mobile host reconnects.

The Oracle Lite only supports flat transactions at mobile hosts. When the mobile host connects to the database server, a refresh process will be performed to synchronize the snapshot with the master copy. If the snapshot is modified, the updates will be sent to the database server. All local transactions at the mobile host will be validated at the database server in the same order as they were executed at the mobile host. The refresh process is a blocking process. This means that no database operations will be allowed at the mobile host during the reconciliation process.

IBM DB2 Everyplace

IBM DB2 Everyplace is an architecture that consists of a relational database at mobile hosts and a mid-tier on fixed hosts. The mid-tier system supports data synchronization between the mobile databases that reside at the mobile hosts with the source databases on the fixed database servers. When mobile hosts are disconnected, transactions are processed locally at the mobile hosts. When mobile hosts reconnect to the database server, synchronization processes are carried out to reconcile data. As Microsoft SQL Server CE and Oracle Lite, IBM DB2 Everyplace does not discuss mobile support stations. Data synchronization processes are carried out directly between the mobile hosts and the fixed database servers.

The IBM DB2 Everyplace only supports flat transactions. When the mobile host connects to the database server, a synchronization process will be performed to synchronize data between the mobile hosts and the source database. IBM DB2 Everyplace differentiates the data synchronization processes between the mobile host and the source database. The synchronization request is submitted from the mobile host and placed in the input queue at the fixed database server. If the synchronization request is allowed to proceed, the data at the mobile host is temporarily saved in the Staging table then the Mirror table. If there is any conflict, it will be resolved in the Mirror table. After this, the changes are stored in the DB2 log and sent to the source database through a Change Data table. For the data synchronization from the source database to the mobile host, an inverse process is performed. The main difference between these two data synchronization processes is that the data from the source database is immediately processed and transferred to the mobile host without any delay, i.e., without passing through the Staging table and Administration control.

CONCLUSION

In this chapter, we have reviewed the basic concepts of database systems and database transactions, and discussed the architecture of transaction processing systems in distributed environments. We will shift our focus to transactions and transaction processing in mobile environments, which possess some unique characteristics such as the mobility of mobile computing hosts, the limitations of wireless communications and the resource constraints of mobile computing devices. We will investigate two important topics: How the distinguishing characteristics of the mobile environments impact transactions and transaction processing systems; and what new requirements a transaction processing system must have in order to efficiently support transaction processing in the mobile environments? In next chapter gives the brief knowledge about the secure mobile payment transaction system in real worlds.

REFERENCES

Maassen, Nieuwpoort, Veldema, Bal, & Plaat. (1999). An efficient implementation of Java's Remote Method Invocation. *Proceedings of the 7th ACM SIGPLAN Symposium on Principles and Practice of Parallel Programming.*

Martin, Bergan, & Russ. (1988). Parpc: A system for parallel procedure calls. *Proceedings of the International Conference on Parallel Processing*, 449-452.

McCann & Roman. (1999). Modeling mobile IP in mobile UNITY. *ACM Transactions on Software Engineering and Methodology, 8*(2).

Mogul, Rashid, & Accetta. (1987). The Packet Filter: An efficient mechanism for user-level network code. *Proceedings of the 11ᵗʰ ACM Symposium on Operating Systems Principles.*

Montenegro & Drach. (1995). System isolation and network fast-fail capability in Solaris. *Proceedings of the Second USENIX Symposium on Mobile & Location-Independent Computing.*

Mummert, Ebling, & Satyanarayanan. (1995). Exploiting weak connectivity for mobile file access. *Proceedings of the 15th ACM Symposium on Operating Systems Principles.*

Myers. (1997). *SMTP service extension for authentication.* Retrieved from http://globecom.net/ietf/draft/draft-myers-smtp-auth-05.shtml

Noble, Price, & Satyanarayanan. (1995). A programming interface for application-aware adaptation in mobile computing. *Proceedings of the Second USENIX Symposium on Mobile & Location-Independent Computing.*

Noble, Satyanarayanan, Narayanan, Tilton, Flinn, & Walker. (1997). Agile application-aware adaptation for mobility. *Proceedings of the 16th ACM Symposium on Operating Systems Principles.*

Object Management Group, Inc. (1997). *CORBA Services: Common Object Services Specification.* Author.

Ott, Michelitsch, Reininger, & Welling. (1998). An architecture for adaptive QoS and its application to multimedia systems design. *Computer Communications, 21*(4), 334–349.

Chapter 5
Secure Mobile Transactions (M–Payment)

ABSTRACT

The Payment Proxy Server monitors and analyses the content headers in HTML and WML pages moving between the content providers and merchants. Whenever it intercepts priced content it initiates a payment transaction and redirects the user to the Payment Server. Pricing and provisioning of the digital content and the payment service can be done through the administration tools. These tools are also used to monitor the events occurring in the payment server. In this chapter, we describe a system that achieves a secure data transmission over a mobile voice channel with a further goal to provide secure voice transmission. Here we describe each component of the system in detail and discuss the issues that we encountered while building the system.

INTRODUCTION

From the consumer's perspective, on-line banking presents numerous benefits to humans, like instantaneous get entry to money owed and balances capacity to conduct remote banking transactions and investments, and finishing touch of electronic programs. With online banking, time and site end up mistaken on situation that those contributions could be accessed at any time; no matter anywhere the man or woman is found. The prospect of round-the clock get

DOI: 10.4018/978-1-5225-2759-6.ch005

access to financial organization assistance and additionally the convenience of transacting business from anywhere within the worldwide have to be specially attractive to shopper, on condition that the flexibleness that e-banking allows appears to fit our increasingly more compartment existence. on line banking makes coinciding service to diverse customers with various goals attainable. Online banking less difficult and convenient for users both historical structures and service. Online banking assists banks of their transition from multiple places to a remunerative and global market. What is attractive for banks to increase growth toward online banking is that the income that they gain by lowering the personnel charges and conjointly generation driven expenses that creates a massive distinction as compared to the traditional banking.

REQUIREMENTS FOR ONLINE TRANSACTION

Mobile Phones

Mobile phones (Small and Seltzer 1996) used in this experiment are two Samsung Galaxy SII phones Nokia N950, Nokia Xpress Music 5530 and Sony Erycsson Xperia Pro. All of these models have ports for 3.5 mm 4-conductor TRRS phone connectors (hands-free headset connector). N950 and Xperia Pro both have an option to enable noise suppression, whereas Xpress Music and both Galaxy SII phones do not. We have upgraded an operation system of one of the Galaxy SII phones to enable noise suppression option.

In the course of experimentation, we have discovered that the XpressMusic would add background noise to the signal and interpret some input signal as "end call" command. For this reason, we did not use Xpress Music afterwards. To enforce a client-side encryption of a voice channel, a microphone captures the voice (A) before it reaches the mobile phone. The voice signal is then redirected into an encryption & modulation unit. An audio codec processes the analog voice signal into digital data (B). Digital data are then encrypted (C). Encrypted digital data are modulated into an analog signal (D). The data-carrying analog signal is sent to a mobile phone (E). Steps (F) and (G) follow as in a usual mobile call. In case of a digital data transmission, the voice recording and encoding steps are omitted and the process follows steps C-D-E-F-G. Of the experiments, we used a Galaxy SII as a sending device and either N950 or Xperia Pro as a receiver.

Encryption and Modulation Unit

In our prototype, an encryption and modulation unit is a Dell Latitude E5410 notebook computer. It has a 3.5 mm headphones output (headset-out) and a 3.5 mm microphone input (microphone-in) ports. It also has an internal microphone. The computer runs Debian GNU/Linux distribution with kernel version 3.11-2. 3.1.3 Custom Headset to connect a mobile phone via its headset port to a microphone-in and headset out ports of a computer, we modified a standard hands-free headset. A typical headset cable usually provides two channels for an audio output from a mobile phone (left and right headphone), one auxiliary input for a microphone and a grounding pin. In the modified cable, an audio output goes into a 3.5 mm connector on the other end of the cable. Another 3.5 mm connector in connected to the auxiliary pin of the headset connector instead of a microphone. Although all four phones have a headset port and recognize both headphones and a microphone with their stock headset, only Galaxy SII recognized the microphone on our custom headset.

Input Processing

In our experiments, we used three types of input data: a plain text, an encrypted text and a pre-recorded voice signal. To ensure that the transmission link is correctly setup, we first used a plain text as an input. If the text was recognizable on the receiver end, the encrypted text and encoded audio were sent.

Voice signal is pre-recorded as one 16-bit channel of PCM encoded audio data with a sample rate of 8000 Hz. Audio data are then compressed using a speech codec codec2 with a bit rate parameter set to 1200. An encryption and decryption utility crypt was used for encryption. Crypt's prominent advantage over other alternatives (such as crypt and Open-SSL) is its capability to decrypt. Alternatively, a Caesar cipher was used for byte-by-byte encryption.

Transmitter

Modulation

The next step in our pipeline after input processing is signal modulation. Based on an analysis of different modulation methods and available software solutions (Alan Snyder 1987), we used a modulation implementation in a

FDMDV modem. By default, the FDMDV modem uses a narrower frequency band of a carrier signal than that of human voice. Default settings place 14 carrier signals around a center frequency of 1500 Hz with 75 Hz separation between carriers. This results in a modulated signal in a frequency band of 900-2050 Hz and a data transmission rate of 1400 bps.

We modified the source code of the modem to use 20 carriers. The resulted modulated signal was in a frequency band of 700-2300 Hz. The data transfer rate of modified modulator was 2000 bps. On the left is a plot spectrum of a signal created by an FDMDV modem with default settings. 14 data signals are multiplexed into one signal with frequency band of 900-2050 Hz. Two noticeably higher peaks in the center represent a clock signal of the modem, which has higher energy than data signals. Seven data signals are on each side of this center. On the right, the FDMDV modem multiplexed ten data signals onto each side of the center frequency. The multiplexed signal is in a frequency range of 700-2300 Hz. We could have increased the data rate even more by filling the voice band with carriers and decreasing distance between them. The bottleneck in our experiments was, however, not the throughput of the channel, but a high error rate of the connection caused by channel noise and signal impairments.

Noise Suppression

After the modem has modulated data into a sound signal, we redirect in into the phone. At this point, the mobile phone can perform noise suppression. Although there was no noticeable effect on the input signal, the experiments showed that the error rate of transmission was lower up to ten percent with noise suppression enabled on the transmitting phone.

Receiver

In order to recover initial data from a signal that came over mobile voice channel has to be performed in a reversed order. The incoming signal is passed from the phone through the modified headset and into the computer. Then the computer performs signal demodulation and data decryption. In case of a voice transmission, digital data represent an encoded sound, which needs to be decoded and played back.

Recording Sound From Mobile Phone

The signal that has originated from a sending side has passed through a mobile voice channel into the receiving mobile phone and is redirected via headphones output into the computer. We discovered two types of signal impairments here. In our experiments, a constant white noise would appear in the recorded signal, even if no device was attached to the microphone-in port of the computer. After some troubleshooting, two separate problems were pinpointed. A GNU/Linux kernel bug was a cause to some of the noise and was fixed by a patch. Another cause of the noise was introduced by the internal microphone which happened to pick up noises made by the notebook itself. The noise would disappear after we unplugged notebook's charger and all peripheral devices.

Intermediation Distortion

Intermediation distortion is a signal integrity issue that arises when multiple signals are sent over a single transmission channel at the same time. The separate signals will mix or multiply with each other. The product of this mixing is called a signal harmonic. Harmonics are quieter than the original signal, but will mix and multiply again with the original signal and other harmonics. This creates a distinct pattern in a plot spectrum of the received signal. Intermediation distortion is a major issue in systems where one medium is used by both transmitter and receiver. Examples of such systems include cellular base stations, duplex radio and wireless systems and satellite systems.

After we canceled out or prevented the noise caused by computer software and hardware, the error rates of most transmissions were still too high for us to understand the originating text or sound. Although the received modulated signal sounded like the original, the plot spectrums of two signals were different.

The pattern in plot spectrum diagrams remained the same with 14- and 20- carrier signals. Further research showed that the same kind of distortion appears in a usual mobile conversation. This signal impairment looks like it was caused by intermediation. A deference of frequencies between human voice and human voice sent over mobile channel.

Demodulation

After recording the modulated signal from a mobile phone, we pass it to the demodulator. The output of the demodulator is a stream of bits. We noted that this stream differs from original data. Firstly, the data stream differs as a result of the demodulator trying to extract data from silence from the beginning of the input signal. Also, the beginning of the Figure 3.5: Intermediation Distortion. 3rd and 5th order harmonics of transmitting signal create a distinctive noise pattern in a plot spectrum of the signal. In a mobile voice channel, harmonics of the transmitting signal overlap with the receiving signal. With voice, this creates insignificant distortion, with modulated data; the noise causes issues when demodulating the signal. Data stream can be scrambled due to the calibration process of the demodulator. Secondly, the demodulated data will have some incorrect bits as a result of mobile voice channel distortion. The negative effect of the mobile channel was heavier than we expected, due to the intermediation distortion. Finally, the amount of data will be bigger, because the demodulator will try to extract data from the silence at the end of the signal.

Post-Processing

The goal of a post-processing step is to restore original data from the stream of bits that comes from the demodulator output (Talukdar, Badrinath, and Acharya 1998). In case of a voice transmission, demodulated data represent an audio encoded by speech codec. The data has to be decoded and can be played back after that. It is not significant that has more data at the beginning and at the end than the original, it will sound as a short sequence of noise. The errors throughout the file are, on the other hand, an issue for the decoder and the player. The higher the error rate gets, the harder it will be to comprehend the extracted voice. Multiplex signal created by the FDMDV modem is noticeably distorted. The received signal contains noise on frequencies that were empty on the transmitting side. For comparison with the original signal, Mobile voice channel distortion of human voice. On the left is a plot spectrum of a human voice. On the right is a plot spectrum of a human voice received from mobile voice channel. Alternatively, if the original data were an encrypted text, it has to be decrypted. Since we know that the demodulated data is corrupted, we instruct the decryption utility to ignore mismatched keys and attempt the decryption.

CONCEPTS FOR MOBILE PAYMENT

As mobile devices have been transforming into personal trust devices, mobile payment is recognized as interactions between parties in a e-payment system with specific context (e.g. business models, player relationships) and capabilities (mobile device capabilities) so that there is at least one party as a mobile user. Basically, the context of m-payments includes any payment in which a mobile device is used in order to "initiate, activate, and confirm" the payment. The m-payment services are often carried out through a none-bank party (such as financial and credit institutions) independent of preexisting bank accounts. Mobile payment systems evolve with new technologies, since they are free of limitations usually applied to bank-anchored services. There are three initiatives that could be considered to best suit mobile payments. First, a mobile device is the most convenient and possible payment technology for mobile context and service purchases. Second, the diminishing use of cash provides the potentials to develop new substitute payment approaches for low value transactions using financial service stations. Third, need of a cost-effective means to charge macro-payments in m-commerce environment. M-payment system is merely registering and forwarding the authorized and validated payment transactions. Payment system life-cycle includes payment request creation, payment request authorization, and payment request committal.

Principally, m-payments occur between four stakeholders: mobile consumers subscribe to a service, merchants, who provide product or service to consumers, payment service provider, which controls the payment process and the trusted third party that administers the authentication of other players and the authorization of payment settlement. Note that different roles can be merged into one party and act as one player. For example, payment service provider, which controls payment process and trusted third party, can act as the same stakeholder.

- **Bluetooth:** Bluetooth, as a popular short-range communication technology, enables mobile devices to communicate with each other using 2.45 GHz frequency at the distance of up to 80 meters, although distances of up to 10 meters are more common. BT traditionally supports data transfer rates up to 3 Mbps, though BT 3.0, which incorporates 802.11 standards (Terry, Theimer, Petersen, Demeres, Spreitzer, and Hauser 1995), can support transfers even up to 24 Mbps. Also, Bluetooth technology provides a better connection, since data transfer dispatches signals in all directions. The impressing aspect of Bluetooth is its impact on increasing development of peer-to-peer use for Bluetooth devices. In addition to file-sharing tasks, a more common task is using Bluetooth-enabled devices to interact with other BT-enabled devices in the intermediate proximity. Other benefits of BT communication protocol is that the messages can be sent to other phones without knowing phone number. These messages could later be used or shared with other users or exchanged for commercial purposed, like exchanging digital goods, tickets or coupons.

Mobile Payment Models

As already explained, transaction of digital value basically includes three phases: payment initiation, payment authorization, and payment settlement. Mobile payment models can be characterized based on some important features, such as: payment amount, payment settlement mechanism, and the technologies which support the complete m-payment system.

With respect to monetary value of a payment, the substance of value will be digital or paper cash. Digital cash can be used as equivalent to the paper cash. Basically it preserves user's anonymity and mainly enables off-line transactions. While in other models payment value should be verified by third-party operators.

Considering the payment amount, the payment will be either macro or micro payment. A macro-payment usually involves amounts more than $10 especially for credit card payments. A typical micro payment scenario basically will be settled by third-parties for authorization and verification request by card issuing bank or financial institutions and in other words, banks pay mobile merchant for the user. While micro-payments normally deal with less than $10 amounts and usually are charging users facilitated by mobile network operator through the billing system.

Mobile payments can be classified into three major types in terms of payment settlement mechanisms: Account-based payment systems, which are based on mobile phone numbers, smart card or credit cards. In account-based systems, the transaction amount is charged by the mobile subscriber's account or credit/debit card. Generally, account-based payment model includes four parties: customer, merchant, issuer or customer's financial service provider, and acquirer or merchant financial institution. In some occasions, there might be another party called payment gateway or proxy acting as an interface between issuers and acquirers in the network of banking side and customer and merchant at the Internet side. In this payment system, each user owns a specific account. With respect to the type of the supporting technology, m-payments are classified into contactless and remote. 'Contactless' or 'proximity' payments are performed with the physical presence of a customer at the point of sale and actually 'face-to-face' or 'machine-to machine'. For example: buying a product from a vending machine. Contactless payments use a radio-frequency interface between the mobile device and the beneficiary's payment device. In fact, remote mode transactions are performed over the network or 'over-their' (OTA). OTA services utilize mobile device facilities that transmit data via GPRS, 3G or Wi-Fi. Using OTA m-payment, consumers can initiate transactions directly from their devices to the payment service provider. In the OTA mode, the confidentiality and integrity of data will be satisfied due to strong encryption and integrity circumstances. Contactless or proximity payments can be fully or partially initiated or settled 'over-the-counter' (OTC) according to available proximity communication facilities in mobile device.

Mobile Payment Transactions

As already explained, there are two main classes of mobile payment systems with different natures of mobile payment transactions. In OTC payments, the customer is physically present at the point-of-sale, and mainly the transaction is conducted using a wireless device using proximity communication protocols. While, in OTA payments, payment transactions are performed where the consumer is physically remote from the point-of-sale and closer to Internet payment gateways. OTA payments usually require a more sophisticated infrastructure for wide acceptance of payment requests. To complete a mobile payment transaction, three steps must be successfully performed in sequence: payment request creation, payment request authorization and, payment request settlement.

Mobile Payment Applications

Mobile payment applications (Thekkath and Levy 1994) can basically an interface handling financial functions which allows a user to perform various payments between the customer account, merchants and bank and also to purchase goods, coupon and offers by different merchants using Smart15 phones. Hence, mobile payment applications are mainly required to provide a secure route for payment request from customers to merchants and vice versa when being used as POS on a merchant side.

Various security services must be provided for mobile payment services. Also, there are specific threats for mobile payment services. There are previous research results about providing essential security services and probable threads corresponding to the payment services occasions. On the other hand, based on different deployment methods of financial mobile applications, there are different architectures of payment methods. There are some common issues in mobile payment standards.

- **Security:** Considers the usage and trust of customers and merchants in the integrity of the payment network, it is vital to increase security level by applying security services: confidentiality, authentication, integrity, authorization, availability and none-repudiation.
- **Interoperability:** It is preferred that any typical payment method can be used at any participating mobile commerce system.
- **Usability:** It is required to consider users' consumption behavior and habits to make the payment system more user-friendly.
- **Privacy:** It is required to protect collected information of the participants of transactions took place over the Internet or stored locally on client side. This information can be useful outside of the transactions, so that any of that information may be linked in a way the participants without their knowledge or consent. Privacy requirement prevents use or disclosure of any personal information and keeps them secure according to defined privacy policy and obligations.

In order to design desired payment options, some issues such as regional support, consumer preferences and customer base should be considered. So, the desired goal of a typical m-commerce system is to establish a balanced trade-off between main issues of mobile payment standards for mobile users to perform convenient e-commerce transactions in a simple and secure way.

For developing markets there are many security concerns for m-payment systems. For instance, there are mobile applications environments that include kind of security model, where mobile payment parties connect to each other in two independent secure approaches. Actually, depending on m-payment architecture, there might be one secure connection either between service provider and mobile operator or between mobile operator and user's mobile terminal or even a secure connection between user's mobile terminals.

Hence, to assure a secure transaction, there is an implication that every secure transaction involving its entities (service providers, mobile peers and mobile operator) requires a trust level between all entities. All of transaction players should be sure that its connection is extended in the secure way to other transaction participants. Also, if the content of connection is being encrypted and decrypted by transaction peers, that may make it vulnerable to potential threats for each transaction entity. So, end-to-end authentication is a desired feature. Moreover, in developing markets, m-payment service providers rely on agents for customer acquisition and payment verification. They use customer's sensitive information and credentials for identification and authentication purposes. These agents are vulnerable to be compromised to customers' information leakage. Furthermore, mobile devices are potentially vulnerable with malwares performing unauthorized operations, such as sending sensitive user information using available connectivity features.

In the following, the most relevant research results are described in order to emphasize different aspects of payment scenarios and security arrangements corresponding to security threats and vulnerabilities, as well as interoperability and usability. In general, these payment protocols comprise transaction parities including customer, merchant, agents, service provider, and mobile operator network. Then, for all the discussed payment methods, the security issues of payment messages will be reviewed and a candidate payment protocol with transaction messages security will be proposed as this thesis purposes and potential security threats will be discussed as well as a supporting potential future work.

ARCHITECTURE FOR SECURE TWO-PARTY MOBILE PAYMENT

The involving parties are a customer and a payment service provider. This architecture focuses on applying security at the application layer. They considered the application layer's security isolated from security protocols in the lower layers. The architecture is designed in such a way that it handles all the security-related functions free of any modifications to the existing communication infrastructure and protocols.

They proposed a secure architecture for two-party mobile payment based on application-layer security architecture to provide end-to-end security, implementing a digital signature module.

Security Mechanism

During a payment transaction, the system transfers the transaction message attached with the digital signature's public key over an unsecured network link. In order to protect transaction messages from third party eavesdropping, both signature and encryption layers are used to process messages. In this architecture, digital signature layer ensures that the message is sent from the right client to the right server. Hence, they combined the SIM (Subscriber Identifier Module Number), PHID (mobile phone serial number) and ACCID (user's bank account number) as the Client ID, then signed all that and appended to the message. Client → Bank: Encrypt {Sign {message, Client ID}}

Due to the J2ME limitations, ECDSA has been adopted because of its low computational cost, higher performance, a fast signature generation, and short key size. Elliptic Curve Digital Signature Algorithm (ECDSA) is adopted to implement Digital Signature Algorithm (DSA). In this architecture, both of two key pairs are used for encryption and digital signature generation. As Figure 5 illustrates, private key generated on a mobile device is used for generating digital signature while other party uses the corresponding public key to verify the digital signature. To keep the key pair (including private key) secret, they are generated and stored in the mobile device.

Basically, when Java applications are being compiled, class files are generated in machine language so; this process makes it difficult to understand details of the private key. The RMS (Record Management System) APIs provides the ability to manipulate records between different applications and shares records within an application, so that access to these records is strictly prohibited. The key pair will eventually expire, and the banking server detects if any renewal of the key-pair in needed then, initiates the renewal of a key pair by notifying mobile device to generate a new one.

A Lightweight and Secure Protocol for Mobile Payments Via Wireless

Internet in M-Commerce

Mentioned some major problems, as follows:

1. Participation of the merchant and acquirer in wallet payment protocol may raise processing rates and make the payment process to take longer.

2.   Cryptography algorithms, hash and digital signatures have been highly used.

3.   Limiting users to use particular devices supporting special software.

In addition to solving these problems, they proposed a new method making the security mechanism simpler.

1.   Initially, the customer initiates the payment using her mobile device by sending a request to her issuer. This is basically withdrawal of money from his/her account and transfer to merchant's account. This request contains the following information: account number or credit card number from which customer needs to withdraw, a complex key (being generated from users' secrets) and, the destination account or credit card number along with the amount of money.

2.   Then, the customer can commit the transaction through a payment gateway provided by the issuer. All the information being transferred via gateway between the issuer and customer, is encrypted/decrypted with WTLS protocol.

3.   Issuer performs all transactions between himself and the acquirer in a secured tunnel.

4.   Depending on some established agreements, acquirer sends a text message or an email to merchant after transferring money.

Issuer rolls back all these transactions in case of any problem during any phase of transactions.

During payment process, the customer generates a complex key using a private key given by the issuer and a number associated with him. This complex key in spite of its simplicity is different for each payment transaction. So, only the issuer and the customer know about its details and how it is been generated, so that it guarantees authentication and non-repudiation. Also, during payment process acquirer needs to create payment authorization response. A complex key will be generated. When this information is transferred to the issuer having it checked for their validity, commit or roll back the transaction. Actually, there is an agreement established between the issuer and acquirer about the algorithms. This agreement can be used for generating the complex using agreed algorithms. In all steps of exchanging data, the complex key is being used for cryptography to provide confidentiality and integrity.

Secure Mechanism Based on Concurrent Signature for Mobile Payment Services

This research has been performed a review over a mobile payment system and its security issues and proposes a payment model and protocol based on concurrent signature scheme. The research earlier evaluates a payment model then, proposes a payment protocol to resolve probable weakness points. The payment model is the following:

Initially, client starts (Waldo, Wyant, Wollrath, and Kendall. 1994) the payment from the wallet application to the payment gateway (PG); next, PG exchanges messages with associated banks and merchants and sends the result back to clients. Actually, Wallet using the security module embedded in the application software performs the encryption of all exchanged messages. Also, all users having mobile access interfaces can connect to a CA for the assignment of an authentication key. In the above model, the security of the model is based on PKI. All entities of the payment service have their own digital certificates.

In the above model, the merchant and the customer need to trust each other. So, an exchange protocol is required to ensure that no one can take an advantage over the other party by misbehaving the protocol. PKI could be an alternative scheme for establishing fair information exchange. But, the relatively high cost of PKI for mobile devices with low computing power caused to compensate this. So, an algorithm named "Concurrent Signature" excluding any third party has been presented to decrease the complexity especially for mobile payments' computations. To resolve problems mentioned above, a mobile payment protocol based on Concurrent Signature has been proposed, involving only clients and merchants, which guarantees the confidentiality, authenticity, non-repudiation.

Security Mechanism

Details of the exchange protocol for mobile services using "Concurrent Signature". The details of the message content, hash and encryption functions can be found in.

As the client/Alice chooses a digital product by her mobile devices, she receives the shopping order. Then, the order will be encrypted by the Alice's signature, attached to it is an unsigned check, and will be sent to merchant/Bob.

Merchant receives the Client's message and verifies of all received messages in terms of validity. Based on the verification result, Bob sends the hint messages back to Client. Next, Merchant chooses a secret key k to encrypted digital services and then computes hash of k (f=Hash(k)) and encrypts the message, sales service commitment and f by k. Merchant encrypts k by public key of Client and sends back all these messages to Client. Client decrypts messages received from Merchant by its private key and retrieves the k, then decrypts service messages to retrieve message content, sales service commitment and f. Client checks the sales service commitment and its signatures check by f and if it was satisfied with the result, it sends S back to Merchant otherwise; it rolls back the whole transaction:

S= (w, h, f, Client's public key, Merchant's public key, C) in which f stands for Hash(k) and h for a cryptographic hash function and w stands for a hash function over Client's private key. When Merchant receives message S, it performs signature check and further validation. If validation process was successfully validated confirmed, the check is confirmed. Then, merchant sends shared secret k to Client, and sends the signature check attached with k to the corresponding bank servers for value transfer. Bank, after successful validation of the legitimacy of Merchant as the receiver of check by k, performs the transfer. Otherwise, it rejects the transfer request. Finally, Client decrypts message using key k which has been received from the Merchant and gets the digital goods.

SYSTEM DESIGN AND ARCHITECTURE

In this thesis, different mobile payment systems have been considered and evaluated relevant to our proposed system design. There are two groups of criteria that ought to be considered relevant: functional and architectural. The functional criteria basically should enforce the system policy and what the system should be able to do to satisfy the system requirements listed and the architectural criteria, i.e. Interoperability, Usability, Simplicity, Security, Privacy, Trust, Cost and, Availability define how the system should be constructed.

Functionality

The purpose of this thesis is to construct a system that enables payment using available Wi-Fi, GPRS, 3G, and Bluetooth compliant mobile terminals which are equipped with mobile devices. The following criteria are being fulfilled by the system in terms of its functionality:

1. Implementing a secure means of initiation, authorization and, settlement of payments, using either credit or bank accounts.
2. Implementing an integrated and secure registration of consumers.
3. There are three actors that interact with the payment system: the customer, the merchant, and the agent.
4. Providing an authentication service to authenticate the end-users in efficient ways in a flexible manner by any available and required means and also, to provide protection for data exchange and authorization.
5. Enabling customers to perform payment transactions directly with the payment system and also, should enable merchants to register payment transaction requests as well as agents to confirm mobile payment transactions, with details about the payment transaction into the payment system.
6. Keeping the record of payment information including registration status of users (customers, merchants and agents) and the status of payment transaction requests and payment messages received by the system.
7. Providing the consistency and integrity of payment transactions initiated by consumers and committed into the payment system.
8. Committing the registered payment transactions, only when the payment scenario has been successfully performed and all engaged consumers have authorized the transaction and the authorization is verified by the system.

Security

The system's design in terms of security should be in line and compatible with other involved components of the system. Hence, the system design and implementation is dependant to security modules and issues related to key lengths, key generation, certificate issuance, distribution and, revocation and, security module implementations. All these issues are considered in order to fulfill the following required security requirements and accomplish the desired performance of the payment system:

1. Provisioning mobile payment applications including preparing and loading mobile applications into user's mobile device with personalized keys and also deployment of unique personalized keys to protect information store and retrieval and the transactions made by the m-applications, so that, user's profile and data exchange must be secured to ensure that no data is compromised. This functional security requirement can provide the security service of "non-repudiation".
2. The keys should be generated compatible with key generation formats and standards of other modules and components of the system. Also, key generation functions should store and retrieve generated keys from hardware security modules installed in mobile devices, that they cannot be recovered by any means.
3. All mobile applications should follow PKI standards defined by the CA of public key certificates.
4. All configuration files and data should be encrypted in secure elements of mobile devices using compatible and standard encryption modules of the system.
5. Access to the system and the system database has to be provided only through the specific interfaces provided by the system.
6. Access to the files and the mobile application database has to be provided only through the specific interfaces provided by the application.
7. All communications between mobile applications and the back-end gateways and servers must be encrypted.
8. OTC transactions should be performed only by authorized users and mobile applications.

System Architecture

The preliminary research effort led me to propose an architectural design for Secure Mobile Payment System with certain design considerations in order to provide and satisfy recently introduced requirements for mobile commerce and financial transactions. This architecture should comprise components, protocols and interfaces to provide various services to various mobile applications: registration, security of users at different levels, and protection of its own components. The architecture is required to overcome the following challenges:

Interoperability, Usability, Simplicity, Security, Privacy, Trust, Cost, Availability, and Cross border payments in order to operate as widely adopted mobile payment architecture.

The system architecture is designed in a modular way so that, it is possible to plug new services and components compatible with the system standards into the system without interfering with other modules, services and components. Basically, the proposed architecture design is based on a regulatory environment where every mobile payment transaction goes through the bank accounts of payment involved parties in flow form of consumer-bank-consumer. The mobile devices mainly act as an interface to access the bank through back-end payment gateways systems. For the design of a system for secure mobile payment, there are some important security aspects in mobile environments which should be considered. In order to provide efficient security, it is required to secure all participating components of the architecture, as well as communication between these components. One of the main components of the system architecture is an infrastructure for support and settlement of financial transactions. A service oriented security infrastructure is adopted suitable for mobile financial transaction. In this infrastructure, a system exists which is called SAFE (Secure Applications for Financial Environment) which is designed and implemented to provide a secure, convenient and reliable large-scale infrastructure for mobile financial transactions.

The components of this system architecture are secure mobile applications for different kind of consumers (Agent, Merchant and, Customer), Location-based authentication service and, three *SAFE Servers*: Communications (Gateway) Server, IDMS (Identity Management System) Server, and Payment Server. All these components are integrated through a secure messaging system and can provide a number of secure financial services and for this thesis specifically payment services.

This chapter mainly presents necessary design and guidelines for implementation of the system. According to requirements and criteria specified in previous chapters, a high level modular model for the system will be specified to describe the functionality and the processes that the system impalements.

As already introduced, the ultimate purpose of this thesis is proposing a system for secure mobile payment transaction using a new method to provide efficient security for payment information messages using existing mobile payment models. In order to design desired payment options, some issues, such as regional support, consumer preferences and customer base, should be considered. So, one of the desired goals of a typical m-commerce system is to provide a balanced trade-off between main issues of mobile payment standards for mobile users to perform convenient e-commerce transactions in a simple, and secure way anytime and anywhere.

System Components

The components of this system architecture are secure mobile applications for different kinds of consumers (Agent, Merchant and, Customer), Location-based authentication service and, three SAFE servers: Communications (Gateway) Server, IDMS (Identity Management System) Server, and Payment Server comprising the SAFE system. All these components are integrated with secure messaging system and can provide a number of secure financial services and for this thesis specifically payment services.

SAFE SYSTEM

SAFE (Secure Applications for Financial Environment) is a system capable of performing various financial transactions with mobile clients or any kinds of mobile devices. SAFE system supports transactions with multiple banks, direct client-to-merchant payments, peer-to peer transactions, and other non-banking mobile applications.

 To follow specifications of the system design, account-based payment model is adopted as one of the payment models supported by the SAFE system. As already explained, this model includes four parties: customer, merchant, issuer or customer's financial service provider, and acquirer or merchant financial institution. In the SAFE model, as the main features of the system, there are mobile pre-paid accounts (PPAs) used to deposit and withdraw cash and also for various mobile payments, so that a consumer can pay with an account associated with its mobile phone number through communication networks. When the customer intends to make an m-payment transaction, he/she can access this wallet, select from which account they want to pay and the beneficiary account number. Then the value is debited from the account of the customer and is transferred to the merchant account (Watson 1994).

 SAFE system can provide the following services based on the system design:

1. **Management and Registration of Pre-Paid Accounts (PPAs) for System Actors:** Agents, Customers and business entities (content providers and merchants);
2. **Use of Those PPAs for Financial Transactions:** over-the-counter (OTC) and over-the-air (OTA) payments, cash deposits/withdrawals, and account transfers.
3. Issuance and management of biometrics smart cards for system administrators, SAFE agents, customers, and merchants for authentication, authorization and payment against PPAs.

SAFE system supports various types of transactions. In this system architecture design, mobile banking and mobile commerce are the main financial applications for payment transactions.

Mobile banking, at an advent of mobile payment technologies, refers to provision and settlement of financial services using mobile devices. SAFE system, offers mobile banking services with its business model, actors and use cases. SAFE mobile banking service can be said to comprise three concepts:

1. Mobile accounting
2. Mobile brokerage
3. Mobile financial information services

Mobile accounting and mobile brokerage can provide transaction-based use cases while non transaction-based use case can be considered as mobile financial information services vital for conducting transactions such as, balance inquiries which might be required prior to credit transfer. All these services are provided in combination with information services which in contrast, maybe provided as an autonomous unit.

The participants or actors in m-Banking transactions are the following:

*   **Banks:** Perform registration and certification of individuals and provision of financial services,
*   **Consumers:** Individuals initiating or receiving transfers as the result of financial transactions to accomplish the described SAFE mobile banking model, large-scale, federated security architecture is designed which comprises two general types of servers in a mobile banking system:

- ○ **Gateway Servers:** Specialized servers that support various secure communication functions, used as the front-end proxies to bank servers.
- ○ **Bank Servers:** Internal servers in banks, performing standard banking applications and transactions.

Meanwhile, client mobile stations including mobile devices enhanced with secure applications and used as an interface interacting with the system to perform financial transactions from mobile locations.

In the following, money transfer transaction is illustrated:

"Money Transfer" transaction may be performed between two personal accounts or between a personal and a corporate account. In any case, one customer is the sender or initiator of the transaction and the other customer is the recipient. This transaction as a primary and basic transfer function may be used for all kinds of payments and money transfer affairs. If the two parties of money transfer have accounts in the same bank, then the sender initiates the transfer of certain amount of money indicating his/her account and the account of the recipient.

Transfer_Request message, including all required parameters, is sent from the sender's mobile application to the SAFE servers, which upon successful verification and commission of transfer by the bank, informs the recipient about the transaction result status. Using the SAFE system, users can perform m-commerce transactions either using over-the- Counter (OTC) or Over-the-Air (OTA) protocols. For OTC transactions, users basically use proximity protocols supported by available their mobile devices (Bluetooth or NFC), while merchants use specialized PoS terminals or mobile devices capable of performing proximity protocols. For OTA transactions, users can use their mobile devices containing secure wallet applications in order to perform full transfer cycle through wireless protocols. Moreover, secure wallet application can also perform standard debit/credit card payments. All the information of the credit cards can be entered into customer's mobile device either during registration or during the process equivalent to credit cards issuance and then stored in a secure element of a mobile device through the provided interfaces of the wallet application. Merchant PoS terminals and mobile devices containing wallet application will be capable to accept such data through proximity protocol. Wallet application user interface is efficiently designed to follow the same steps in today's debit/credit cards transactions.

Payment using smart card in OTC mode includes using mobile device to reach merchant PoS terminal or mobile device to provide cards number and other data for merchant through available proximity protocols. While in OTA mode, initiating payment by selecting an already registered card will be possible through wireless protocols. While merchant needs to connect to SAFE server to verify the authorization of transaction and upon receiving authorization result and termination of transaction, merchant sends back the transaction result.

Also, "Digital Cash Dispensing and Micropayments" are supported by SAFE system. In fact, it is possible to load cash from recently registered bank accounts or credit cards to your mobile device and digitally store it. So, instead of cash you can use stored "digital cash" for later micro-payment transactions. Hence, you can use the type of payment using corresponding mobile devices equipped with hardware and software supporting appropriate proximity protocol and required mobile application.

In order to load digital cash into customer's mobile device, customer needs to send "Cash_Request" message via available interfaces through mobile application to the SAFE server. Then, after successful validation, "digital cash" is debited from customer's account, loaded and stored in his/her mobile wallet. When the customer needs to initiate a micropayment using OTC mode, he/she initiates the payment using proximity tools to PoS side or merchant mobile device then the payment amount is reduced form customer's "digital wallet" and transferred to merchant PoS terminal or mobile device. Finally merchant sends "Cash_Reclaim" message to the SAFE server which contains merchant's bank account number to make deposit into it. For OTA mode, customer simply can choose its pre-loaded digital wallet as the source account for payment and the rest of transaction goes on the same as with OTC mode through wireless protocols. Moreover, it is possible to unload digital cash from digital wallet back to any registered bank account or credit card.

Mobile Shopping

Mobile shopping feature of SAFE system encapsulates the following functionalities to cover new e-commerce trends.

M-Promotions

M-Promotions is a functionality that is been considered in our system design to make a convenient way for merchants to present information to customers as well as others and increase demands of their products and services and also for customers to receive the latest popular special offers on their mobile device. Basically, promotions can be uploaded by merchant through available interface on their corresponding mobile application available on the market visible for customers through their wallet applications. Customers can select and download the available promotions filtered by user's location into their wallet application, so that they can view stored promotions all in wallet offline mode.

M-Coupons

M-Coupons is a functionality that is been considered to provide digital coupons containing financial discounts or rebate issued by manufacturers of consumer package goods and services, to be used in retail stores as part of promotions. M-coupons functionality not only can enable prices conscious customers to use this coupons and a form of price discrimination, but also, can enable merchants to offer coupon with lower price offers, targeted selectively to regional markets specifically where regional markets with great price competitions. SAFE system has enabled merchants to upload their coupons into m-marketing server using provided function of the merchant smart device application. Also, customers can view available coupons in the market filtered by customer's location and select those that have been uploaded by merchants close to the customer's location and store them in wallet application.

M-Tickets

M-Tickets function enables customers to inquire, pay for, obtain and validate tickets conveniently using mobile devices. This functionality can reduce the production and distribution costs of traditional paper-based tickets by providing simple ways to purchase tickets conveniently. Moreover, this functionality provides merchant to upload any kind of ticket to the SAFE system and customers, on the other side, can view available tickets in the m-market server of SAFE system via designed corresponding interfaces of mobile application. These tickets can be purchased and then downloaded into wallet application. Tickets on the m–Marketing Server can be filtered by customer's location. Tickets may be paid using OTC/OTA payment method – SAFE accounts, bank accounts, bankcards, or digital cash and upon successful payment be downloaded and stored in wallet application.

M-Parking

M-Parking provides four functionalities:

1. **Search Parking:** Users may search available parking place at the local garage / parking location by giving its registration number
2. **Pay Parking:** Users may pay for parking by specifying parking place, parking time and selecting method of payment.
3. **Extend Parking:** The system will warn users about expiration of their parking time. In that case, users may extend parking by specifying parking place number and additional parking time.
4. **Pay Ticket:** Users may use SAFE system to pay tickets for parking violations. The system will notify users if the ticket has been issued for parking violation. In that case, users may pay the ticket using SAFE system, by specifying ticket number, parking place number, amount to pay and by selecting the method for payment.

All mobile parking functions require SAFE system to be integrated with the parking system of some Parking Authority.

M-Gift Cards

M-Gift Cards functionality enables merchants to upload their gift-cards as a restricted monetary equivalent issued by retailers or banks as an alternative to non-monetary gifts to SAFE m-Marketing server via mobile application interfaces. Customers can list gift cards loaded into the m–Marketing Server. These gift cards can be purchased using OTC and OTA payment methods and then downloaded and stored in the wallet application. Gift card can be transferred to another SAFE user and may be used as one of the payment options with designed payment interfaces in wallet mobile application.

Mobile Applications

SAFE Wallet application supports three groups of mobile functions:

1. Various types of mobile payments using mobile pre–paid accounts, standard bank accounts, bankcards and digital cash payments in OTA and OTC mode. Wallet application provides an interface for customers to perform various types of payment using different types of accounts that SAFE system supports. Using Wallet application, customers can use a specific type of payment based on the situation and available capability of host mobile device.

2. Wallet application enables its user to register and create associated pre-paid SAFE accounts, as well as credit/debit card registration in order to provide most of the usual payment methods.

3. Users can perform deposit, withdraw and transfer transactions using their pre-paid SAFE mobile accounts between all SAFE account owners and then list all launched transactions filtered by defined constraints.

SAFE Merchant application supports three basic mobile functions:

1. Mobile payments with subscribers using over–the–air (OTA) SMS and wireless protocols. Using merchant application, OTC payment is performed using Bluetooth and NFC proximity protocols and OTA using wireless protocol, either provided by the Internet providers or telecom carriers.

2. Various types of mobile marketing functions: creating and uploading promotions, mobile coupons, mobile gift–cards, and mobile tickets into the SAFE system marketing server.

3. Various types of mobile business functions: accepting discounts based on promotions, mobile coupons, mobile gift–cards, and verification of mobile tickets. Also, providing mobile transactions at locations using PoS systems, then mobile services for the owners/drivers of vehicles, and various mobile commerce transactions and accepting and clearing mobile vouchers from customers, uploaded by merchants using m–Marketing services.

4. Mobile security functions: registration of a merchant and administration of and access to SAFE mobile pre–paid accounts, configuration of the SAFE system, detection of locations, management of Merchant's local

and SAFE system PINs, selection of cryptographic options, and managing X.509 digital certificates.

SAFE Agent application is being used by agents, merchants or any other member of the SAFE system, authorized by the system administrator to perform the functions of an agent.

1.   Self–registration and registration of subscribers and merchants, including registration of locations for merchants
2.   Cash–in and cash–out financial transactions with subscribers and merchants.
3.   Mobile security functions: registration of an agent, management of agent's local and SAFE system PINs, selection of cryptographic options, and managing X.509 digital certificates.

SECURITY COMPONENTS AND ARCHITECTURE

The goal of this research is designing a secure system for mobile transactions. Also, the system design and architecture support various financial mobile applications and transactions. Since all the functions and transactions are basically financial operations, the main concern must be their security. Therefore, one of the most distinguished features of the whole system architecture is its comprehensive security. In fact, system architecture is designed in such a way that existing components can be enhanced with security countermeasures, so that the integrity and availability of the whole system would be preserved. Here, the security infrastructure of the system architecture will be described. Fundamentally, according to the inherent characteristics of financial transactions, the item of trust should be established between participating parties. There are security requirements that must be supplied in the security architecture design to provide the sufficient trust:

Authentication

User identification verification and approval is essentially required in system design and must be efficiently provided. Each participant in the system, including all functions and transactions, needs to make sure that counterparty is the one he/she is interested to communicate with. There are some factors as the "basic instruments available to a human user to authenticate her in order to convince a computing system of her true "identity", as is known or registered in the system". These factors are called authentication factors and are generally classified into three categories:

1. **What an Entity Knows:** It is something secret that ideally only the valid subject should know, for example, password, PIN, answers to security questions.
2. **What an Entity Has:** Identity cards and licenses or any physical tokens which imply identities may make entities recognizable. These factors are usually something physical that the user owns it and only the user who owns the correct token can be successfully authenticated.
3. **What an Entity Is:** basically based on physical characteristics of the user which are uniquely associated with him/her, such as fingerprint, the pattern of user's voice or face.

Integrity

In this context specifically means that transaction contents must be created or modified only by authorized parties or only in authorized ways to assure all participants that the received messages have not been altered in any way from the original message. Generally a message digest of the original message is attached with the message for the recipient to verify the integrity. The cryptography prevents changing the data block (the plaintext) and also changing the checksum value (the cipher-text) to match.

Confidentiality

Confidentiality ensures that the transaction contents are accessed only by authorized parties. Basically access can be reading, viewing, printing or knowing that a particular asset exists. In this context, encryption and decryption are the methods to achieve confidentiality. Both types of crypto systems are used to provide confidentiality: symmetric and asymmetric cryptography.

Non-Repudiation

As a security requirement it provides and maintains evidence, so that the participants of transactions or interactions cannot deny their participation in that transaction. There are some factors required to provide non-repudiation qualified enough as one of the system security requirements:

1. Capture information about the actions that a participant did in transactions or system events. Here, the information is required which are explicit enough to help assign accountability.
2. Preserve and protect all information required to achieve non-repudiation associated with an event. It is quite important to keep all non-repudiation data uncorrupted in order to enforce accountability. Increasing non-repudiation service quality provides a degree of confidence about integrity and reliability of information.
3. Preserve availability of system services in addition to non-repudiation service. Non-repudiation service should be accomplished so that system participants get discouraged due to transaction complexity and duration.

Privacy

This is the function of protecting the collected information of the participants of transactions that were performed over the Internet. This information can be useful outside of the transactions, so that any of that information may be linked in a way the participants without their knowledge or consent. Privacy requirement prevents use or disclosure of any personal information and keeps them secure according to defined privacy policy and obligations.

Availability

As an integral requirement, each security infrastructure needs the complete security design in order to provide expected services available enough for its intended users. Security design architecture provides necessary protection for all data both on a client side (mobile device) and at a server side to achieve an end-to-end security. The following components comprise design of the proposed security architecture:

1. **Local Security Module:** Providing local (mobile device) security using encryption procedures and secure elements in order to protect sensitive local data stored in a mobile device. Identity Provider (Registration) Servers: providing registration services to create maintain and manage of identities and their associated information. Certification Servers: issuance, management and distribution of X.509 digital certificates for system participants based on PKI.

2. **Authorization Servers:** Providing authentication and authorization services for system authorities based on secure web services and, for consumers, based on location-based authentication services.

Users establish transaction flows both with other users and with various service providers via SAFE Gateway Server. The SAFE Gateway Server, as the core component of the SAFE system, provides communication with users at the front-end through any available kind of wireless communication protocols and connects with security providers and service providers at the back-end through stable TCP/IP connections. It receives various requests from clients in SAFE messages format and upon interpretation dispatches these request messages to different requested service providers. Security providers basically provide security services synchronized with service providers' needs and client-side standards, so that they can accomplish the most efficient access control and transaction security countermeasures. They are playing key role in providing security services needed to accomplish security requirements. Communication protocol between security providers, service providers, and front-end components are all standardized and designed to achieve efficient system functionality.

CLIENT SECURITY MODULE

Front-end side clients communicate with the system using smart mobile devices. In fact, users may communicate with back-end servers through various mobile applications. Mobile device client internal structure comprises four independent modules. In other words, mobile device includes mobile application and removable secure element. SAFE Mobile application logically comprises:

1. **Communication Module:** For establishing communications with other nodes using available and supported communication protocols and protecting communication using application layer security circumstances.
2. **Business Logic Module:** Containing business logic of the application function and transaction logics and procedures.
3. **Security Module:** Providing required security capabilities required for application functions and transaction messages security as well as communication module.

Secure element being used with the described mobile applications, are removable SD cards. SE basically provides a secure environment for storing sensitive information and cryptographic keys.

IDMS SERVER

IDMS server provides services for registration of SAFE system participants and their authentication, authorization, roles and privileges within the system boundaries in order to increase security and productivity of the entire security infrastructure. Registration of users may be performed either via registration interfaces in mobile applications or by SAFE Agents under supervision of a bank, a telecom operator or any other independent ID services provider, so that all the system entities have reliable and verifiable registration data used for all SAFE transactions. IDMS server in our security architecture provides directory and access control services supporting identity management and single sign-on for other components of security infrastructure.

CA SERVER

CA Server provides management of digital certificates of the system entities. Since in our security architecture transaction security is based on PKI, CA server performs all the activities related to digital certificates. CA server issues digital certificates to the owners of public keys generated by the SAFE mobile application and stored in secure elements of mobile devices. CA server issues X.509 certificates based on registration data provided by IDMS server. Digital certificates contain public key generated in customers' mobile devices using specified key generation functions. Generated public key and private key will be stored in secure element of the mobile device through provided interfaces of the SAFE mobile applications. Customers can request their digital certificates using mobile application functions upon supplement of required information associated with their generated public key. Then, after successful issuance of certificates by the CA server, they may be stored in mobile phones secure elements. After reliable and verifiable registration, certification, and issuance of smart cards, an instance of the SAFE system is ready to support various secure financial transactions.

SECURE ELECTRONIC PAYMENT

The theory of public key cryptography is discussed in detail, as both, the previous work done in the area of electronic payments in general and the theory related to information security, especially public key infrastructures, as it is the fundamental basis for securing any transactions in a distributed environment.

Mobile Public Key Infrastructure

This chapter presents the emerging wap public key infrastructure (WPKI) specified by the WAP Forum and the MeT. In order to discuss the WPKI, the fundamentals of public key infrastructures in general are also presented. The most widely deployed PKI specification, the X.509, is discussed in a more detail as it is the basis for the WPKI.

Public Key Cryptography

Traditional cryptography is based on a secret, i.e. a key that can be used to both encipher and decipher information. This of course implies that all parties who want to do either operation have to know the secret. Public key cryptography, on which public key infrastructures are based, in turn makes use of two keys one that is kept secret or private key and one that is made public and thus known to everyone. Furthermore the keys are such, that it is computationally virtually impossible to deduce the private key by just having knowledge of the public one. In public key cryptography data enciphered with one of the keys, can only be deciphered using the other.

Since then it has been the mainstream in strong cryptography as it solves the problem of key management that is the major problem in more traditional symmetric cryptography. Public key cryptography can be used to either to accomplish confidentiality or digital signatures. The prior can be achieved when the enciphering key is made public and the deciphering key is kept secret. Signatures, on the other, hand are achieved when the enciphering key is kept secret and the deciphering key is made public.

Digital Signatures

A digital signature, the electronic counterpart of handwritten signatures, possesses the following properties: it is authentic, un-forgeable, non-reusable and non-reputable. Furthermore, the signed document is unalterable after the signature.

Enciphering a document that is the subject to signing with the secret key, creates digital signature. Others can verify the signature by deciphering the enciphered document with the public key. In practical implementations, to save time, complete documents aren't encrypted just for the purpose of signing, because of the inefficiency of existing public key cryptography algorithm. One-way hash functions are used to reduce the documents to be signed to digests that can be efficiently enciphered. The verification of the signatures produced in this manner, involves reproducing the digest and comparing it to the deciphered original digest. Blind signatures are a special form of digital signature. They are of great interest especially in electronic cash systems. A blind signature is accomplished, when the signer doesn't have knowledge of what he is actually signing. A blind signature protocol between A and B works as follows: A selects a random blinding factor that he uses to multiply the message and passes the blinded message to B for signing. B signs the blinded message and passes the signature to A. A divides the signature by the blinding factor to remove, which removes the blinding factor from the signature. A prerequisite of course is that the blinding and signature operations are commutative.

Digital Certificates

Digital signatures by themselves are not sufficient means for automatic verification of authorities – even if a signature can be verified, there is no guarantee of the fact that the person who made the signature is who he claims to be. Public-key certificates are a powerful means of establishing trust in public-key cryptography. A public-key certificate is someone's public key, signed by a trustworthy person.

A public-key certificate normally also contains information related to the secret key related to the signed public key. The trustworthy party naturally signs this extra information along with the key. If the public-key certificate contains information about the identity of the owner of the secret key, the certificate is called an *Identity certificate*. If the attached information attached identifies the

144

key holder as a member of a group or an organization, the certificate is called an *accreditation certificate*. An adaptation of the accreditation certificates are *authorization* and *permission certificates*.

The *requesting party*, *issuing party* and *the verifying party or parties:* The issuer is normally called a *certification authority* (CA). The requesting party must perform a proof of possession of the secret key, and to supply sufficient information about his identity that can physically be verified to the CA. Certification authorities can form so called certification or trust chains by certifying each other. Thus the trustworthiness of a CA in a chain depends on that of all the CAs before it in the chain. The first CA the one that is implicitly trusted – is called the root certification authority.

X.509 Public Key Infrastructure: X.509 is an ITU-T recommendation. Its core is the use public-key identity certificates with each user of a system. The certificates in X.509 are identity based. The recommendation specifies the certificate format as well as the role of the CA. The CAs can create hierarchical trust models - i.e. all but the root-CAs are certified by other CAs. Also cross certification is possible. This enables creating trust relationships between different CA certification trees. The X.509 is based on a X.500 directory. According to the paradigm there is a global directory, where there is an entry for each individual. The certificates are stored in the X.500 directory along with other data about the individuals. According to the original approach, the names of the individuals would be globally unique. In real world deployments of the X.509 the directories normally cover only one organization or e.g. the users of some application. In practice there can't be any strict uniqueness requirements for the names in the X.500 directory, at least across different directories. Thus, also the certificates need, often be considered as a permission to use some application or service, rather than a 100% proof of identity. Normally this is an adequate level of authorization, since the whole notion of certification relies on the issuance process - the users are positively at one point as members of the group and they are issued the certificates.

Centralized Secret Key Management

In a centralized secret key management scheme, the key pairs are generated centrally on a key server. The private key is stored permanently on the server and it never leaves it. The key holders have access credentials to the server, which they can use to access their key, and perform signing and encryption operations. The 3D SET, specified by Visa and some other credit card companies makes use of the centralized scheme. It is an extension of the normal SET, which was based on a distributed scheme. The problem with the distributed SET scheme, that is secret keys to be stored on the PCs of the users and special client software has to install in order to use the SET system. The fundamental problem of a centralized secret key management scheme is that doesn't give the key holders real control over their secret keys. Depending on the authentication scheme used to access the key, the level of security reached can vary greatly; it may even be non-existent, in case a user id and a badly selected password are used.

Distributed Secret Key Management Using ICCs

Integrated circuit cards, ICCs are a viable technology for storing secret keys. The ICC is a credit-card like piece of plastic, with a small microchip cast inside it. On top of the card there are the connectors for the chip pins. In a public key cryptography application making use of ICCs, the key pair is normally generated inside the ICC. The card communicates the public key out, but the secret key is kept inside it. The card performs the cryptography operations that need the use of the secret key. Typically there is a PIN code authentication for using the secret key. The advantage of embedding the secret key into the chip in a smart card is that the key cannot be extracted from the card by any means – the card is said to be tamper resistant.

Electronic Payments

Forms of, at least partially, electronic payments have been around since late 1970s [6]. As a communications infrastructure is a prerequisite for electronic payments, the enormous expansion of the Internet has enabled a fast progress in the area during the 1990s. Electronic payment systems are classified in to four groups in:

1. Credit card based systems
2. Electronic checks and Account Transfers
3. Electronic cash payment systems
4. Micropayment systems

The grouping is based for the most part on the business model behind the payment solutions. The grouping may not be the best, in terms of the functional perception of the systems but it serves the purpose of presenting the different types nicely. Differences between the different groups of payment systems are not in all cases very dramatic and in many cases a real life system is hard to classify to belong to any one of the groups for example some of the micropayment schemes could be used to implement e.g. electronic checks.

Credit Card Payments

Credit cards have payments set against an account with a pre-agreed repayment scheme. The entities involved in a credit card transaction are illustrated in the figure below – the scheme is inherent to all account based schemes, and is thus useful to understand. Both the credit card issuers and the acquirers are members of a card association, e.g. Visa or MasterCard. Normally both of them are banks. The merchants have association with the acquirer, who provides the necessary infrastructure for the merchant for accepting credit card payments.

The general schema doesn't change even if the transactions were direct debits from a bank account or a pre-paid account. Payments using credit cards can be made where they are physically presented and e.g. over telephone or Internet i.e. when the card holder and the merchant aren't collocated. This is called a MOTO payment transaction. Clearly this form of payments using credit cards increases the risk borne by the merchants – the merchant doesn't have any means to positively prove the transaction actually took place if the cardholder decides to repudiate it afterwards.

The *Secure Electronic Transaction*, SET, protocols were developed to overcome the possibilities of fraud in credit card transactions over the Internet. It defines two major interfaces – the one between the cardholder and the merchant and that between the merchant and the acquirer (called the payment gateway in the SET context). SET relies on a certification hierarchy, based on the X.509 PKI. Original SET makes use of client software installed on each cardholder's PCs. The software is primarily used to perform digital signing and authentication operations as well as the key and certificate management. As the key problem with SET has been the accumulation of the critical mass of cardholders willing to install the client software another version of SET has been developed. The Three Domain SET is based on having the SET Wallets on a centralized server. The Wallet Server that is accessed with a web browser can perform the same operations as the original client software while removing the requirement of client installation. On the other hand the model compromises security by weakening the level of cardholder authentication.

Electronic Checks and Account Transfers

An authorization to transfer funds from the payer's account to the payee's account. Also the process of handling the checks is similar with both paper and electronic ones – optimally banks can make use of their existing clearing and settlement infrastructure, in case of private electronic check systems, a different scheme may be exercised:

1. The payer authorizes the transfer/payment by writing and signing the check
2. The payee' delivers the check to its bank for clearing and settlement
3. The banks use a clearing and settlement network to transfer the funds

Most of the electronic payment systems on the Internet rely on a centralized account model, where both the payer and the merchant need to have accounts in the same institution and funds can thus be transferred directly from one account to another. Frameworks that support inter-payment provider transfers have also been defined. The most prominent one of those is the FSTC initiatives. The FSTC has defined a format for electronic checks called the FSML. The FSML has an XML like syntax for describing financial documents – especially checks. The FSTC has also defined a number of different functional flows in electronic check payment in the inter-bank case that enable flexible implementation of the actual payment systems.

In centralized account scenario, the most problematic issues are naturally the ways for funding and debiting the accounts. From the payer's perspective, funding the account is normally done by transferring funds from a bank account or by a credit card to the account at the payment provider. In any case, some other form of electronic payment is normally needed for funding the account. The payment provider normally provides APIs for the merchants for debiting the payers' accounts on-line. The merchants normally get the funds out of their accounts with the payment provider via traditional automated account transfers.

Electronic Cash Payment Systems

Cash is the most popular form of payment in retail transactions between consumers and businesses – depending on country 75% - 95% transactions are paid in cash. The attractive properties of cash as form of payment are:

1. Acceptability – cash is almost always accepted
2. Guaranteed payment – cash is safe for the merchant to accept. If the money isn't forged the payment is always honored
3. No transaction charges
4. Anonymity

Many electronic cash payment systems are more or less focusing on a subset of the above. The key advantage of electronic cash over physical cash is the lack of running handling costs. The downside is the fact there isn't a well-established global infrastructure for electronic cash. Clearly electronic cash systems also attract attempts at fraud, someone may try to counterfeit tokens of electronic money, issued electronic money may be attempted to be 'double-spent', i.e. spent many times. Criminals might also abuse electronic cash payment systems for tax evasion, bribes or money laundering, especially when the blind signatures are used. As the goal of this thesis is not to implement electronic money, this type of payment systems is not discussed in more detail.

Micropayment Systems

In applications, where the number of transactions between each payer and the merchant is large and the amount of the each individual transaction is low, the transaction processing cost grows proportionally large. Traditionally this kind of setting is addressed by a subscription scheme the use of a service is paid for on an estimate basis; a bulk amount is paid for which the use of a service is bought for a certain period of time. In practice there is, however, a clear need to be able to pay per use as some people might only occasionally use the service that some use very frequently. Micropayment systems are targeting this problematic.

There are many different micropayment schemes. A group of micropayment systems are based on probabilistic or computational cost of 'minting' payment tokens. E.g. Pay Word, Micro Mint and the probability-based schemes all relax security while reducing processing costs – small-scale fraud may even be relatively easy. Other micropayment schemes aim at providing fool proof secure payments while also reducing the transaction costs, possibly at the cost of flexibility and elegance.

A very representative example of a micropayment scheme belonging to the latter group is jalda. It is a proprietary standard for electronic micropayments specified by the EHPT a joint venture between Ericsson and Hewlett Packard. Jalda is a protocol and an API for implementing charging for service or goods. It is based on a concept of a payment session that is initiated by the payer by accepting and electronically signing a session contract with the merchant. A jalda payment provider verifies the contract for the vendor. After the contract has been verified by the payment provider, the vendor can start sending 'ticks' indicating that the service is being consumed and the payer should be charged. Clearly the protocol is intended for uses like multimedia services, where the user subscribes to e.g. a video stream and he is invoiced during the video plays 'tick' at a time by the service vendor. The protocol makes no constraints on what channel the actual service is consumed over. The protocol is also flexible in terms of the way the digital signature is created.

Ericsson MCP

The Ericsson Mobile Commerce Platform (MCP) is based on two products the Mobile e-Pay and the EHPT Safe Trader.

The Mobile e-Pay is a front-end product for authentication of mobile subscribers. It works over SMS and WAP. In SMS mode the authentication of the users is based on a PIN code, that the user has to submit to the server or optionally a customized SIM Toolkit solution that enables the use of PKI based strong authentication over SMS. With WAP, currently only the PIN code authentication over the network is permitted – in future releases also WAP 1.2 compliant WTLS class 3 and sign Text based methods are said to be supported. When no signing capabilities exist in the client and a PIN authentication is used, the Mobile e-Pay server performs the signing on behalf of the client.

The EHPT Safe Trader is a back-end payment provider system. It is in no way limited to the mobile channel. In practice the Safe Trader is a user account management system. It is intended to be used as the core of a payment provider infrastructure. There are interfaces for all the necessary external systems, like invoicing, pre-payment, revenue assurance and business intelligence. The interfaces are, however, rudimentary and based on file transfer of data.

The fact that the product is based on the proprietary jalda as the integration protocol towards merchant systems doesn't pose problems, as there are publicly available APIs offered free of charge for major programming languages. The inherent nature of jalda, however, forces the merchant to adopt a payment session concept, which may not be perfectly optimal for all purposes. However, they do not depend on each other in any way. When put together, the Ericsson MCP is a quite versatile one. The problem with the package is, that is also pretty expensive considering the problems it solves quite a lot of integration is required to set up a complete system with all the links to the necessary supporting systems. The Mobile e-Pay is a solution in some ways similar to the one specified in this thesis. However, in the Mobile e-Pay support for MeT or WAP 1.2 compliance is only promised in future releases.

Nokia Payment Solution

The Nokia Payment Solution enables payments in mobile terminals. Its focus is on payments for electronic content delivered from web or mobile portals, although the Payment Server can be also used for payments for other goods and services. The solution consists of five major components: the Nokia Payment Server, Nokia Payment Proxy Server, Nokia Virtual Purse application and two administrative tools. The payment concept is based on consuming credits from the pre-paid account managed in the virtual purse application. The payments are authorized using digital signatures created on a WAP 1.2 -based WIM.

The Payment Proxy Server monitors and analyses the content headers in HTML and WML pages moving between the content providers and merchants. Whenever it intercepts priced content it initiates a payment transaction and redirects the user to the Payment Server. Pricing and provisioning of the digital content and the payment service can be done through the administration tools. These tools are also used to monitor the events occurring in the payment server. The Nokia Payment Server manages the interfaces towards financial institutions and e.g. telecommunications billing systems. The solution can thus be integrated to the whole of the enterprise infrastructure.

The Nokia Payment Solution suite is a comprehensive set of services that can readily be used to establish a mobile payment service. At the package was studied for the purposes of this thesis, it was, however, still in development phase and e.g. the WIM-based digital signing capabilities What comes to the system structure of the Nokia Payment Server, it resembles in functionality that of the Mobile Payment System presented in this paper. The concept of automated on-line payment of digital content is, however, not embraced in the Mobile Payment System as it is in the Nokia suite.

CONCLUSION

This chapter describes a system that achieves a secure data transmission over a mobile voice channel with a further goal to provide secure voice transmission. Here we describe each component of the system in detail and discuss the issues that we encountered while building the system and also discussed the secure payment transmission through mobile devices.

REFERENCES

Chandramohan, Kkath, & Levy. (1994). Hardware and software program support for efficient exception managing. *Proceedings of the Sixth International Conference on Architectural Support for Programming Languages and Operating Systems.*

Small & Seltzer. (1996). A comparison of OS extension technology. *Proceedings of the USENIX Annual Technical Conference.*

Snyder. (1987). Inheritance and the improvement of encapsulated software program components. InShriver, B., & Wegner, P. (Eds.), *Research Directions in Object-Oriented Programming* (pp. 165–188). MIT Press.

Talukdar, Badrinath, & Acharya. (1998). Rate model schemes in networks with cellular hosts. *Proceedings of the 4th ACM/IEEE International Conference on Mobile Computing and Networking.*

Terry, Theimer, Petersen, Demeres, Spreitzer, & Hauser. (1995). Managing update conflicts in Bayou, a weekly connected replicated garage system. *Proceedings of the Fifteenth ACM Symposium on Operating Systems Principles.*

Waldo, Wyant, Wollrath, & Kendall. (1994). *A word on distributed computing, Technical Report SMLI TR-ninety four-29.* Mountain View, CA: Sun Microsystems Laboratories, Inc.

Watson. (1994). Application design for wireless computing. *IEEE Workshop on Mobile Computing Systems and Applications.*

Related Readings

To continue IGI Global's long-standing tradition of advancing innovation through emerging research, please find below a compiled list of recommended IGI Global book chapters and journal articles in the areas of banking, finance, and economics. These related readings will provide additional information and guidance to further enrich your knowledge and assist you with your own research.

Abdin, J. (2015). Foreign Direct Investment (FDI) in Bangladesh: Trends, Challenges, and Recommendations. *International Journal of Sustainable Economies Management*, 4(2), 36–45. doi:10.4018/IJSEM.2015040104

Agwu, E. M., & Murray, P. J. (2015). Empirical Study of Barriers to Electronic Commerce Uptake by SMEs in Developing Economies. *International Journal of Innovation in the Digital Economy*, 6(2), 1–19. doi:10.4018/ijide.2015040101

Al-Hasan, S., Thomas, B., & Mansour, A. (2016). Internet Adoption and International Marketing in the Jordanian Banking Sector. *International Journal of Online Marketing*, 6(2), 34–48. doi:10.4018/IJOM.2016040103

Alemu, T., Bandyopadhyay, T., & Negash, S. (2015). Electronic Payment Adoption in the Banking Sector of Low-Income Countries. *International Journal of Information Systems in the Service Sector*, 7(4), 27–47. doi:10.4018/IJISSS.2015100102

Alganer, Y., & Yılmaz, G. (2015). Fiscal Integration and Harmonization: European Union Integration from Fiscal Perspectives – Objectives, Means, Obstacles, and Politics. In E. Sorhun, Ü. Hacıoğlu, & H. Dinçer (Eds.), *Regional Economic Integration and the Global Financial System* (pp. 49–58). Hershey, PA: IGI Global. doi:10.4018/978-1-4666-7308-3.ch005

Alt, R., Eckert, C., & Puschmann, T. (2015). Network Management and Service Systems: The Case of German and Swiss Banks. *Information Resources Management Journal*, *28*(1), 38–56. doi:10.4018/irmj.2015010103

Anthopoulos, L. G., & Siozos, P. (2015). Can IT Innovation become a Tool against Fiscal Crisis?: Findings from Europe. *International Journal of Public Administration in the Digital Age*, *2*(1), 39–55. doi:10.4018/ijpada.2015010103

Artienwicz, N. (2016). Behavioral Stream in Polish Accounting: Its Relation to Behavioral Finance and the Perspectives for Neuroaccounting Development in Poland. In B. Christiansen & E. Lechman (Eds.), *Neuroeconomics and the Decision-Making Process* (pp. 246–261). Hershey, PA: IGI Global. doi:10.4018/978-1-4666-9989-2.ch013

Asongu, S. A., & Nguena, C. L. (2015). Equitable and Sustainable Development of Foreign Land Acquisitions: Lessons, Policies, and Implications. In E. Osabuohien (Ed.), *Handbook of Research on In-Country Determinants and Implications of Foreign Land Acquisitions* (pp. 1–20). Hershey, PA: IGI Global. doi:10.4018/978-1-4666-7405-9.ch001

Aybars, A. (2014). The Relationship Between Institutional Investment and Earnings Management: Empirical Evidence from Turkey. *International Journal of Corporate Finance and Accounting*, *1*(1), 1–21. doi:10.4018/ijcfa.2014010101

Baber, H. (2016). Risk Mitigation Practices in Banking: A Study of HDFC Bank. *International Journal of Risk and Contingency Management*, *5*(3), 18–32. doi:10.4018/IJRCM.2016070102

Bachrane, M., Khaled, A., El Alami, J., & Hanoune, M. (2016). Investment Location Selection based on Economic Intelligence and Macbeth Decision Aid Model. *Journal of Information Technology Research*, *9*(3), 37–48. doi:10.4018/JITR.2016070103

Backović, N., Milićević, V., & Sofronijevic, A. (2016). Strategic Directions in European Sustainable City Management. In M. Erdoğdu, T. Arun, & I. Ahmad (Eds.), *Handbook of Research on Green Economic Development Initiatives and Strategies* (pp. 147–168). Hershey, PA: IGI Global. doi:10.4018/978-1-5225-0440-5.ch008

Ballas, A., Sykianakis, N., Tzovas, C., & Vassilakopoulos, C. (2014). An Investigation of Greek Firms Compliance to IFRS Mandatory Disclosure Requirements. *International Journal of Corporate Finance and Accounting*, *1*(1), 22–39. doi:10.4018/ijcfa.2014010102

Banerji, D., & Das, R. (2014). Critical Review of Curriculum in Legal Financial Studies in Turkey: Perspectives and Prospects. In N. Baporikar (Ed.), *Handbook of Research on Higher Education in the MENA Region: Policy and Practice* (pp. 102–118). Hershey, PA: IGI Global. doi:10.4018/978-1-4666-6198-1.ch006

Baranowska-Prokop, E., & Sikora, T. (2017). Competitiveness of Polish International New Ventures from Managerial Perspective. In A. Vlachvei, O. Notta, K. Karantininis, & N. Tsounis (Eds.), *Factors Affecting Firm Competitiveness and Performance in the Modern Business World* (pp. 83–107). Hershey, PA: IGI Global. doi:10.4018/978-1-5225-0843-4.ch003

Barat, S. (2016). Importance of Customer Satisfaction in a Community Bank. *International Journal of Innovation in the Digital Economy*, *7*(4), 56–73. doi:10.4018/IJIDE.2016100104

Batrancea, L., Nichita, A., Batrancea, I., & Kirchler, E. (2016). Tax Compliance Behavior: An Upshot of Trust in and Power of Authorities across Europe and MENA. In M. Erdoğdu & B. Christiansen (Eds.), *Handbook of Research on Public Finance in Europe and the MENA Region* (pp. 248–267). Hershey, PA: IGI Global. doi:10.4018/978-1-5225-0053-7.ch012

Bihari, S. C. (2014). When Citi was Found Sleeping. In V. Jham & S. Puri (Eds.), *Cases on Consumer-Centric Marketing Management* (pp. 258–277). Hershey, PA: IGI Global. doi:10.4018/978-1-4666-4357-4.ch021

Bird, R. M. (2016). Transparency, Technology and Taxation. In M. Erdoğdu & B. Christiansen (Eds.), *Handbook of Research on Public Finance in Europe and the MENA Region* (pp. 11–29). Hershey, PA: IGI Global. doi:10.4018/978-1-5225-0053-7.ch002

Bodea, C., Stelian, S., & Mogos, R. (2017). E-Learning Solution for Enhancing Entrepreneurship Competencies in the Service Sector. In I. Hosu & I. Iancu (Eds.), *Digital Entrepreneurship and Global Innovation* (pp. 225–244). Hershey, PA: IGI Global. doi:10.4018/978-1-5225-0953-0.ch011

Boitan, I. A. (2016). Early Warning Tools for Financial System Distress: Current Drawbacks and Future Challenges. In Q. Munir (Ed.), *Handbook of Research on Financial and Banking Crisis Prediction through Early Warning Systems* (pp. 97–114). Hershey, PA. doi:10.4018/978-1-4666-9484-2.ch005

Buggea, E., Castiglione, R., Cerquitelli, T., Grosso, L., Rontini, G., Scolari, A., & Xiang, L. (2014). Internationalization Services for Small and Medium Enterprises: A Case Study. In C. Machado & P. Melo (Eds.), *Effective Human Resources Management in Small and Medium Enterprises: Global Perspectives* (pp. 393–414). Hershey, PA: IGI Global. doi:10.4018/978-1-4666-4731-2.ch019

Cabanda, E., & Domingo, E. C. (2014). A Production Approach to Performance of Banks with Microfinance Operations. *International Journal of Information Systems in the Service Sector*, 6(2), 18–35. doi:10.4018/ijisss.2014040102

Çavuşoğlu, T., & Önal, D. K. (2016). A Panel VAR Analysis of the Shadow Economy in Europe and MENA. In M. Erdoğdu & B. Christiansen (Eds.), *Handbook of Research on Public Finance in Europe and the MENA Region* (pp. 201–220). Hershey, PA: IGI Global. doi:10.4018/978-1-5225-0053-7.ch010

Celik, I. E., Dinçer, H., & Hacioğlu, Ü. (2014). Investment and Development Banks and Strategies in Turkey. In Ü. Hacioğlu & H. Dinçer (Eds.), *Globalization and Governance in the International Political Economy* (pp. 131–140). Hershey, PA: IGI Global. doi:10.4018/978-1-4666-4639-1.ch010

Cheng, P. (2016). Fragile by Design: The Political Origins of Banking Crises and Scarce Credit. *International Journal of Applied Behavioral Economics*, 5(1), 48–52. doi:10.4018/ijabe.2016010103

Choudhury, M. A. (2014). Productivity Analysis in Ethically Induced Financing Environment: A Case Study of Indonesian Islamic Banks. In *Socio-Cybernetic Study of God and the World-System* (pp. 192–218). Hershey, PA: IGI Global. doi:10.4018/978-1-4666-4643-8.ch007

Chytis, E., Filos, J., Tagkas, P., & Rodosthenous, M. (2016). Audit Firms, Deferred Taxation and Financial Reporting: The Case of The Athens Stock Exchange. *International Journal of Corporate Finance and Accounting*, 3(1), 1–21. doi:10.4018/IJCFA.2016010101

Chytis, E., Koumanakos, E., & Goumas, S. (2015). Deferred Tax Positions under the Prism of Financial Crisis and the Effects of a Corporate Tax Reform. *International Journal of Corporate Finance and Accounting*, 2(2), 21–58. doi:10.4018/IJCFA.2015070102

Cinelli, S. A. (2017). The World's Oldest Profession - Now and Then: Disruption of the Commercial Banking Model. In W. Vassallo (Ed.), *Crowdfunding for Sustainable Entrepreneurship and Innovation* (pp. 78–89). Hershey, PA: IGI Global. doi:10.4018/978-1-5225-0568-6.ch005

Ciocoiu, C. N., & Cicea, C. (2015). Development of the Green Economy in Romania: Dimensions, Strengths and Weaknesses. In A. Jean-Vasile, I. Andreea, & T. Adrian (Eds.), *Green Economic Structures in Modern Business and Society* (pp. 161–179). Hershey, PA: IGI Global. doi:10.4018/978-1-4666-8219-1.ch009

Cipolla-Ficarra, F. V., & Alma, J. (2014). Banking Online: Design for a New Credibility. In F. Cipolla-Ficarra (Ed.), *Advanced Research and Trends in New Technologies, Software, Human-Computer Interaction, and Communicability* (pp. 71–82). Hershey, PA: IGI Global. doi:10.4018/978-1-4666-4490-8.ch007

Cossiavelou, V. (2017). ACTA as Media Gatekeeping Factor: The EU Role as Global Negotiator. *International Journal of Interdisciplinary Telecommunications and Networking*, 9(1), 26–37. doi:10.4018/IJITN.2017010103

Cura, S. (2015). The Impact of Sovereign Debt Crisis on the EU Economy: Is This the End of the Dream? In E. Sorhun, Ü. Hacıoğlu, & H. Dinçer (Eds.), *Regional Economic Integration and the Global Financial System* (pp. 1–11). Hershey, PA: IGI Global. doi:10.4018/978-1-4666-7308-3.ch001

da Silva, A., Pletsch, C. S., Klann, R. C., Fasolin, L. B., & Scarpin, J. E. (2015). Influence of International Accounting Convergence on the Level of Earnings Management in both Brazilian and Chilean Companies. In I. Lourenço & M. Major (Eds.), *Standardization of Financial Reporting and Accounting in Latin American Countries* (pp. 195–218). Hershey, PA: IGI Global. doi:10.4018/978-1-4666-8453-9.ch009

Dapontas, D. K. (2016). Developing EWS Models for Contemporary Crises Using Extreme Value Binary Models: The Cases of Eurozone and Argentinian Peso (2014). In Q. Munir (Ed.), *Handbook of Research on Financial and Banking Crisis Prediction through Early Warning Systems* (pp. 332–352). Hershey, PA: IGI Global. doi:10.4018/978-1-4666-9484-2.ch016

Datta, N. (2015). Growth and Knowledge Management Strategy of Indian Commercial Banks: A Non-Parametric Approach. *International Journal of Measurement Technologies and Instrumentation Engineering*, 5(1), 28–45. doi:10.4018/IJMTIE.2015010103

Dima, I. C. (2015). Considerations on the Current State of Strategic Management. In I. Dima (Ed.), *Systemic Approaches to Strategic Management: Examples from the Automotive Industry* (pp. 166–218). Hershey, PA: IGI Global. doi:10.4018/978-1-4666-6481-4.ch008

Dinçer, H., & Hacıoğlu, Ü. (2014). The Competitiveness and Strategies in Global Financial System. In H. Dinçer & Ü. Hacioğlu (Eds.), *Global Strategies in Banking and Finance* (pp. 1–13). Hershey, PA: IGI Global. doi:10.4018/978-1-4666-4635-3.ch001

Ebenezer, E. E., Shi, W., & Mackie, W. E. (2015). Chinese Investments in Africa: Implications for Entrepreneurship. In J. Ofori-Dankwa & K. Omane-Antwi (Eds.), *Comparative Case Studies on Entrepreneurship in Developed and Developing Countries* (pp. 99–109). Hershey, PA: IGI Global. doi:10.4018/978-1-4666-7533-9.ch006

Edwards, J., & Newton, S. (2016). Enhancing Regulatory, Financial, Fiscal Investment Incentives as a Means of Promoting Foreign Direct Investment. In M. Ojo (Ed.), *Analyzing the Relationship between Corporate Social Responsibility and Foreign Direct Investment* (pp. 191–201). Hershey, PA: IGI Global. doi:10.4018/978-1-5225-0305-7.ch013

Eken, M. H., & Kale, S. (2014). Bank Branch Efficiency with DEA. In I. Osman, A. Anouze, & A. Emrouznejad (Eds.), *Handbook of Research on Strategic Performance Management and Measurement Using Data Envelopment Analysis* (pp. 626–667). Hershey, PA: IGI Global. doi:10.4018/978-1-4666-4474-8.ch022

Eken, M. H., Kale, S., & Selimler, H. (2014). Analyzing the Efficiency of European Banks: A DEA-Based Risk and Profitability Approach. In H. Dinçer & Ü. Hacioğlu (Eds.), *Global Strategies in Banking and Finance* (pp. 28–55). Hershey, PA: IGI Global. doi:10.4018/978-1-4666-4635-3.ch003

El Dessouky, N. F. (2016). Corporate Social Responsibility of Public Banking Sector for Sustainable Development: A Comparative Study between Malaysia and Egypt. In M. Al-Shammari & H. Masri (Eds.), *Ethical and Social Perspectives on Global Business Interaction in Emerging Markets* (pp. 52–73). Hershey, PA. doi:10.4018/978-1-4666-9864-2.ch004

El-Firjani, E. R., & Faraj, S. M. (2016). International Accounting Standards: Adoption, Implementation and Challenges. In E. Uchenna, M. Nnadi, S. Tanna, & F. Iyoha (Eds.), *Economics and Political Implications of International Financial Reporting Standards* (pp. 231–250). Hershey, PA: IGI Global. doi:10.4018/978-1-4666-9876-5.ch011

Encinas-Ferrer, C. (2017). Currency Parity and Competitiveness: The Case of Greece. In A. Vlachvei, O. Notta, K. Karantininis, & N. Tsounis (Eds.), *Factors Affecting Firm Competitiveness and Performance in the Modern Business World* (pp. 282–299). Hershey, PA: IGI Global. doi:10.4018/978-1-5225-0843-4.ch010

Engwanda, M. N. (2015). Mobile Banking Adoption in the United States: A Structural Equation Modeling Analysis. *International Journal of E-Services and Mobile Applications*, 7(3), 18–30. doi:10.4018/IJESMA.2015070102

Epler, P., & Ross, R. (2015). Spending Options for Service Delivery Models. In *Models for Effective Service Delivery in Special Education Programs* (pp. 91–112). Hershey, PA: IGI Global. doi:10.4018/978-1-4666-7397-7.ch005

Ertürk, E., Yılmaz, D., & Çetin, I. (2016). Optimum Currency Area Theory and Business Cycle Convergence in EMU: Considering the Sovereign Debt Crisis. In R. Das (Ed.), *Handbook of Research on Global Indicators of Economic and Political Convergence* (pp. 67–91). Hershey, PA: IGI Global. doi:10.4018/978-1-5225-0215-9.ch004

Eryigit, S. B. (2016). Does Trust Matter for Foreign Direct Investment Decisions? In M. Al-Shammari & H. Masri (Eds.), *Ethical and Social Perspectives on Global Business Interaction in Emerging Markets* (pp. 224–239). Hershey, PA: IGI Global. doi:10.4018/978-1-4666-9864-2.ch013

Eshraghi, A. (2014). Fund Manager Overconfidence and Investment Narratives. In R. Hart (Ed.), *Communication and Language Analysis in the Corporate World* (pp. 1–20). Hershey, PA: IGI Global. doi:10.4018/978-1-4666-4999-6.ch001

Espagne, E., & Aglietta, M. (2016). Financing Energy and Low-Carbon Investment in Europe: Public Guarantees and the ECB. In M. Erdoğdu, T. Arun, & I. Ahmad (Eds.), *Handbook of Research on Green Economic Development Initiatives and Strategies* (pp. 132–146). Hershey, PA: IGI Global. doi:10.4018/978-1-5225-0440-5.ch007

Even, A., Parmet, Y., & Erez, L. (2015). Factors that Affect Customers Readiness for Internet-based BI Services. *International Journal of Business Intelligence Research*, 6(1), 30–48. doi:10.4018/IJBIR.2015010103

Feldman, R., Govindaraj, S., Liu, S., & Livnat, J. (2014). Optimal Portfolio Construction Using Qualitative and Quantitative Signals. In R. Hart (Ed.), *Communication and Language Analysis in the Corporate World* (pp. 140–161). Hershey, PA: IGI Global. doi:10.4018/978-1-4666-4999-6.ch009

Fiodendji, K., Kamgnia, B. D., & Tanimoune, N. A. (2014). Inflation and Economic Performance in the CFA Franc Zone: Transmission Channels and Threshold Effects. In P. Schaeffer & E. Kouassi (Eds.), *Econometric Methods for Analyzing Economic Development* (pp. 10–29). Hershey, PA: IGI Global. doi:10.4018/978-1-4666-4329-1.ch002

Gáspár-Szilágyi, S. (2017). Human Rights Conditionality in the EU's Newly Concluded Association Agreements with the Eastern Partners. In C. Akrivopoulou (Ed.), *Defending Human Rights and Democracy in the Era of Globalization* (pp. 50–79). Hershey, PA: IGI Global. doi:10.4018/978-1-5225-0723-9.ch003

Gedikli, A., Erdoğan, S., & Yıldırım, D. Ç. (2015). After The Global Crisis, Is It Globalization or Globalonelization? In Ö. Olgu, H. Dinçer, & Ü. Hacıoğlu (Eds.), *Handbook of Research on Strategic Developments and Regulatory Practice in Global Finance* (pp. 287–307). Hershey, PA: IGI Global. doi:10.4018/978-1-4666-7288-8.ch018

Ghobakhloo, M., Hong, T. S., & Standing, C. (2015). B2B E-Commerce Success among Small and Medium-Sized Enterprises: A Business Network Perspective. *Journal of Organizational and End User Computing, 27*(1), 1–32. doi:10.4018/joeuc.2015010101

Goel, S. (2014). Fraud Detection and Corporate Filings. In R. Hart (Ed.), *Communication and Language Analysis in the Corporate World* (pp. 315–332). Hershey, PA: IGI Global. doi:10.4018/978-1-4666-4999-6.ch018

Gordini, N. (2014). Genetic Algorithms for Small Enterprises Default Prediction: Empirical Evidence from Italy. In P. Vasant (Ed.), *Handbook of Research on Novel Soft Computing Intelligent Algorithms: Theory and Practical Applications* (pp. 258–293). Hershey, PA: IGI Global. doi:10.4018/978-1-4666-4450-2.ch009

Guillet de Monthoux, P., & Statler, M. (2014). Theory U: Rethinking Business as Practical European Philosophy. In O. Gunnlaugson, C. Baron, & M. Cayer (Eds.), *Perspectives on Theory U: Insights from the Field* (pp. 234–243). Hershey, PA: IGI Global. doi:10.4018/978-1-4666-4793-0.ch015

Günaydin, D., Cavlak, H., & Cavlak, N. (2015). Social Exclusion and Poverty: EU 2020 Objectives and Turkey. In Z. Copur (Ed.), *Handbook of Research on Behavioral Finance and Investment Strategies: Decision Making in the Financial Industry* (pp. 170–186). Hershey, PA: IGI Global. doi:10.4018/978-1-4666-7484-4.ch010

Gürcü, M., & Tengilimoğlu, D. (2017). Health Tourism-Based Destination Marketing. In A. Bayraktar & C. Uslay (Eds.), *Strategic Place Branding Methodologies and Theory for Tourist Attraction* (pp. 308–331). Hershey, PA: IGI Global. doi:10.4018/978-1-5225-0579-2.ch015

Hackney, D. D., McPherson, M. Q., Friesner, D., & Correia, C. (2014). On the Social Costs of Bankruptcy: Can the Bankruptcy Abuse Prevention and Consumer Protection Act (BAPCPA) of 2005 be an Effective Policy?. *International Journal of Social Ecology and Sustainable Development*, 5(1), 58–91. doi:10.4018/ijsesd.2014010106

Hasan, I., & Pasiouras, F. (2015). Stress Testing and Bank Efficiency: Evidence from Europe. *International Journal of Corporate Finance and Accounting*, 2(2), 1–20. doi:10.4018/IJCFA.2015070101

Henry, E., & Leone, A. J. (2014). Measuring the Tone of Accounting and Financial Narrative. In R. Hart (Ed.), *Communication and Language Analysis in the Corporate World* (pp. 36–47). Hershey, PA: IGI Global. doi:10.4018/978-1-4666-4999-6.ch003

Hocaoğlu, D. (2017). Challenges in Promoting Cities through Culture within the New Global Economy. In A. Bayraktar & C. Uslay (Eds.), *Global Place Branding Campaigns across Cities, Regions, and Nations* (pp. 229–250). Hershey, PA: IGI Global. doi:10.4018/978-1-5225-0576-1.ch011

Homata, A., Mihiotis, A., & Tzortzaki, A. M. (2017). Franchise Management and the Greek Franchise Industry. In A. Vlachvei, O. Notta, K. Karantininis, & N. Tsounis (Eds.), *Factors Affecting Firm Competitiveness and Performance in the Modern Business World* (pp. 251–281). Hershey, PA: IGI Global. doi:10.4018/978-1-5225-0843-4.ch009

Hu, J., Marques, J., Holt, S., & Camillo, A. A. (2014). Marketing Channels and Supply Chain Management in Contemporary Globalism: E-Commerce Development in China and its Implication for Business. In B. Christiansen, S. Yıldız, & E. Yıldız (Eds.), *Handbook of Research on Effective Marketing in Contemporary Globalism* (pp. 325–334). Hershey, PA: IGI Global. doi:10.4018/978-1-4666-6220-9.ch018

James, S. (2016). The Difficulties of Achieving Successful Tax Reform. In M. Erdoğdu & B. Christiansen (Eds.), *Handbook of Research on Public Finance in Europe and the MENA Region* (pp. 30–47). Hershey, PA: IGI Global. doi:10.4018/978-1-5225-0053-7.ch003

Jean-Vasile, A., & Alecu, A. (2016). Trends and Transformations in European Agricultural Economy, Rural Communities and Food Sustainability in Context of New Common Agricultural Policy (CAP) Reforms. In A. Jean-Vasile (Ed.), *Food Science, Production, and Engineering in Contemporary Economies* (pp. 1–24). Hershey, PA: IGI Global. doi:10.4018/978-1-5225-0341-5.ch001

Jeločnik, M., Zubovic, J., & Djukic, M. (2016). Implications of Globalization on Growing External Debt in Eight Transition Economies. In V. Erokhin (Ed.), *Global Perspectives on Trade Integration and Economies in Transition* (pp. 80–104). Hershey, PA: IGI Global. doi:10.4018/978-1-5225-0451-1.ch005

Jindrichovska, I., & Kubickova, D. (2016). Economic and Political Implications of IFRS Adoption in the Czech Republic. In E. Uchenna, M. Nnadi, S. Tanna, & F. Iyoha (Eds.), *Economics and Political Implications of International Financial Reporting Standards* (pp. 105–133). Hershey, PA: IGI Global. doi:10.4018/978-1-4666-9876-5.ch006

Kablan, A. (2014). Financial Control and Ratio Analysis in Local Governments. In H. Dinçer & Ü. Hacioğlu (Eds.), *Global Strategies in Banking and Finance* (pp. 410–422). Hershey, PA: IGI Global. doi:10.4018/978-1-4666-4635-3.ch027

Karaca, C. (2016). The Comparison of the Shadow Economy in Turkey and European Countries. In B. Christiansen & M. Erdoğdu (Eds.), *Comparative Economics and Regional Development in Turkey* (pp. 73–105). Hershey, PA: IGI Global. doi:10.4018/978-1-4666-8729-5.ch004

Karaibrahimoglu, Y. Z., & Tunç, G. (2014). Financial Statement Analysis under IFRS. In N. Ray & K. Chakraborty (Eds.), *Handbook of Research on Strategic Business Infrastructure Development and Contemporary Issues in Finance* (pp. 238–255). Hershey, PA: IGI Global. doi:10.4018/978-1-4666-5154-8.ch017

Kasemsap, K. (2015). The Role of E-Business Adoption in the Business World. In N. Ray, D. Das, S. Chaudhuri, & A. Ghosh (Eds.), *Strategic Infrastructure Development for Economic Growth and Social Change* (pp. 51–63). Hershey, PA: IGI Global. doi:10.4018/978-1-4666-7470-7.ch005

Kasemsap, K. (2015). The Role of Electronic Commerce in the Global Business Environments. In F. Cipolla-Ficarra (Ed.), *Handbook of Research on Interactive Information Quality in Expanding Social Network Communications* (pp. 304–324). Hershey, PA: IGI Global. doi:10.4018/978-1-4666-7377-9.ch019

Katou, A. A., & Katsouli, E. F. (2017). Empirical Evidence on Convergence of Travel and Tourism Competitiveness and Global Competitiveness Across the BRIC Countries. In M. Dhiman (Ed.), *Opportunities and Challenges for Tourism and Hospitality in the BRIC Nations* (pp. 1–14). Hershey, PA: IGI Global. doi:10.4018/978-1-5225-0708-6.ch001

Korres, G. M., & Kokkinou, A. (2014). Public Spending Efficiency: The Missing Factor through Financial Crisis. *International Journal of Social Ecology and Sustainable Development*, 5(4), 1–10. doi:10.4018/ijsesd.2014100101

Kushwaha, G. S., & Agrawal, S. R. (2015). Customer Management Practices: Multiple Case Studies in Stock Broking Services. *International Journal of Customer Relationship Marketing and Management*, 6(2), 1–14. doi:10.4018/IJCRMM.2015040101

Lara-Rubio, J., Martínez-Fiestas, M., & Cortés-Romero, A. M. (2014). Drop-Out Risk Measurement of E-Banking Customers. In F. Liébana-Cabanillas, F. Muñoz-Leiva, J. Sánchez-Fernández, & M. Martínez-Fiestas (Eds.), *Electronic Payment Systems for Competitive Advantage in E-Commerce* (pp. 143–162). Hershey, PA: IGI Global. doi:10.4018/978-1-4666-5190-6.ch009

LaRocca, R. N. (2014). Assessing the Political and Socio-Economic Impact of Corruption among Nations. *International Journal of Information Systems and Social Change*, 5(4), 18–40. doi:10.4018/ijissc.2014100102

Li, S. (2014). Pre-GFC Bank Behaviour Change and Basel Accords. In *Emerging Trends in Smart Banking: Risk Management Under Basel II and III* (pp. 35–56). Hershey, PA: IGI Global. doi:10.4018/978-1-4666-5950-6.ch003

Liu, L. (2014). Micro-Analysis of the Bank of China. In *International Cross-Listing of Chinese Firms* (pp. 226–233). Hershey, PA: IGI Global. doi:10.4018/978-1-4666-5047-3.ch007

Lokuwaduge, C. S. (2016). Exploring the New Public Management (NPM)-Based Reforms in the Public Sector Accounting: A Sri Lankan Study. In A. Ferreira, G. Azevedo, J. Oliveira, & R. Marques (Eds.), *Global Perspectives on Risk Management and Accounting in the Public Sector* (pp. 49–67). Hershey, PA: IGI Global. doi:10.4018/978-1-4666-9803-1.ch003

Long, P., & Vy, P. D. (2016). Internet Banking Service Quality, Customer Satisfaction and Customer Loyalty: The Case of Vietnam. *International Journal of Strategic Decision Sciences, 7*(1), 1–17. doi:10.4018/IJSDS.2016010101

Lopes, F. C., Morais, M. P., & Sasvari, P. (2014). Comparative Analysis on the Usage of Business Information Systems among Portuguese and Hungarian Small and Medium-Sized Enterprises. In H. Rahman & R. de Sousa (Eds.), *Information Systems and Technology for Organizational Agility, Intelligence, and Resilience* (pp. 265–296). Hershey, PA: IGI Global. doi:10.4018/978-1-4666-5970-4.ch013

Lopez-Iturriaga, F., & Pastor-Sanz, I. (2016). Using Self Organizing Maps for Banking Oversight: The Case of Spanish Savings Banks. In Q. Munir (Ed.), *Handbook of Research on Financial and Banking Crisis Prediction through Early Warning Systems* (pp. 116–140). Hershey, PA: IGI Global. doi:10.4018/978-1-4666-9484-2.ch006

Lu, Y. (2016). Public Financial Information Management for Benefits Maximization: Insights from Organization Theories. *International Journal of Organizational and Collective Intelligence, 6*(3), 50–74. doi:10.4018/IJOCI.2016070104

Man, M. (2015). Budgeting Technique of Strategic Management. In I. Dima (Ed.), *Systemic Approaches to Strategic Management: Examples from the Automotive Industry* (pp. 328–362). Hershey, PA: IGI Global. doi:10.4018/978-1-4666-6481-4.ch012

Marois, T. (2016). State-Owned Banks and Development: Dispelling Mainstream Myths. In M. Erdoğdu & B. Christiansen (Eds.), *Handbook of Research on Comparative Economic Development Perspectives on Europe and the MENA Region* (pp. 52–73). Hershey, PA: IGI Global. doi:10.4018/978-1-4666-9548-1.ch004

Marwah, G. S., & Ladhani, V. (2016). Financial Sector in Afghanistan: Regulatory Challenges in Financial Sector of Afghanistan. In A. Kashyap & A. Tomar (Eds.), *Financial Market Regulations and Legal Challenges in South Asia* (pp. 224–262). Hershey, PA: IGI Global. doi:10.4018/978-1-5225-0004-9.ch011

Medda, F. R., Partridge, C., & Carbonaro, G. (2015). Energy Investment in Smart Cities Unlocking Financial Instruments in Europe. In A. Vesco & F. Ferrero (Eds.), *Handbook of Research on Social, Economic, and Environmental Sustainability in the Development of Smart Cities* (pp. 408–433). Hershey, PA: IGI Global. doi:10.4018/978-1-4666-8282-5.ch019

Mertzanis, C. (2015). Marketing Financial Services and Products in Different Cultural Environments. In B. Rishi (Ed.), *Islamic Perspectives on Marketing and Consumer Behavior: Planning, Implementation, and Control* (pp. 232–267). Hershey, PA: IGI Global. doi:10.4018/978-1-4666-8139-2.ch011

Michael, O. B. (2015). Performance Measurement Systems and Firms' Characteristics: Empirical Evidences from Nigerian Banks. *International Journal of Business Analytics*, 2(3), 67–83. doi:10.4018/IJBAN.2015070105

Milgram-Baleix, J., Parravano, M., & Pedauga, L. E. (2014). The Role of B2B E-Commerce in Market Share: Evidence from Spanish Manufacturing Firms. In F. Liébana-Cabanillas, F. Muñoz-Leiva, J. Sánchez-Fernández, & M. Martínez-Fiestas (Eds.), *Electronic Payment Systems for Competitive Advantage in E-Commerce* (pp. 1–14). Hershey, PA: IGI Global. doi:10.4018/978-1-4666-5190-6.ch001

Mion, L., Georgakopoulos, G., Kalantonis, P., & Eriotis, N. (2014). The Value Relevance of Accounting Information in Times of Crisis: An Empirical Study. *International Journal of Corporate Finance and Accounting*, 1(2), 44–67. doi:10.4018/ijcfa.2014070104

Montero-Romero, T., & Cordobés-Madueño, M. (2014). Enterprise Resource Planning System (ERP) and Other Free Software for Accounting and Financial Management of Non-Profit Entities. In J. Ariza-Montes & A. Lucia-Casademunt (Eds.), *ICT Management in Non-Profit Organizations* (pp. 73–89). Hershey, PA: IGI Global. doi:10.4018/978-1-4666-5974-2.ch005

Mukherjee, S., & Chakraborty, D. (2016). Does Fiscal Policy Influence Per Capita CO_2 Emission?: A Cross Country Empirical Analysis. In S. Dinda (Ed.), *Handbook of Research on Climate Change Impact on Health and Environmental Sustainability* (pp. 568–592). Hershey, PA: IGI Global. doi:10.4018/978-1-4666-8814-8.ch028

Munir, Q., & Kok, S. C. (2016). Early Warning System for Banking Crisis: Causes and Impacts. In Q. Munir (Ed.), *Handbook of Research on Financial and Banking Crisis Prediction through Early Warning Systems* (pp. 1–21). Hershey, PA: IGI Global. doi:10.4018/978-1-4666-9484-2.ch001

Nisha, N. (2016). Exploring the Dimensions of Mobile Banking Service Quality: Implications for the Banking Sector. *International Journal of Business Analytics*, *3*(3), 60–76. doi:10.4018/IJBAN.2016070104

Okon, E. E. (2016). Multinational Enterprises and African Economy. In M. Khan (Ed.), *Multinational Enterprise Management Strategies in Developing Countries* (pp. 351–381). Hershey, PA: IGI Global. doi:10.4018/978-1-5225-0276-0.ch018

Olgu, Ö., & Yılmaz, E. (2014). Foreign Ownership and Bank Efficiency: Evidence from Turkey. In H. Dinçer & Ü. Hacioğlu (Eds.), *Global Strategies in Banking and Finance* (pp. 75–100). Hershey, PA: IGI Global. doi:10.4018/978-1-4666-4635-3.ch006

Özer, A. C., & Gürel, H. (2017). Internet Banking Usage Level of Bankers: A Research on Sampling of Turkey. In S. Aljawarneh (Ed.), *Online Banking Security Measures and Data Protection* (pp. 27–39). Hershey, PA: IGI Global. doi:10.4018/978-1-5225-0864-9.ch002

Öztayşi, B., & Kahraman, C. (2014). Quantification of Corporate Performance Using Fuzzy Analytic Network Process: The Case of E-Commerce. In P. Vasant (Ed.), *Handbook of Research on Novel Soft Computing Intelligent Algorithms: Theory and Practical Applications* (pp. 385–413). Hershey, PA: IGI Global. doi:10.4018/978-1-4666-4450-2.ch013

Pan, B., Wei, S., Xu, X., & Hong, W. (2014). The Impact of Defense Investment on Economic Growth in the Perspective of Time Series: A Case Study of China. *International Journal of Applied Evolutionary Computation*, *5*(4), 44–58. doi:10.4018/IJAEC.2014100104

Patro, C. S., & Raghunath, K. M. (2016). Corporate Social Responsibility: A Manifestation in FDI. In M. Ojo (Ed.), *Analyzing the Relationship between Corporate Social Responsibility and Foreign Direct Investment* (pp. 202–227). Hershey, PA: IGI Global. doi:10.4018/978-1-5225-0305-7.ch014

Peng, E. Y., Shon, J., & Tan, C. (2014). Market Reactions to XBRL-Formatted Financial Information: Empirical Evidence from China. *International Journal of E-Business Research*, *10*(3), 1–17. doi:10.4018/ijebr.2014070101

Plaza i Font, J. P. (2016). The European Union as a Chaotic System. In Ş. Erçetin & H. Bağcı (Eds.), *Handbook of Research on Chaos and Complexity Theory in the Social Sciences* (pp. 33–42). Hershey, PA: IGI Global. doi:10.4018/978-1-5225-0148-0.ch003

Popescu, G. H. (2015). The Reform of EU Economic Governance. In G. Popescu & A. Jean-Vasile (Eds.), *Agricultural Management Strategies in a Changing Economy* (pp. 100–118). Hershey, PA: IGI Global. doi:10.4018/978-1-4666-7521-6.ch005

Puia, G. M., Affholter, J. A., & Potts, M. D. (2015). Factors Stimulating Entrepreneurship: A Comparison of Developed (U.S. and Europe) and Developing (West African) Countries. In J. Ofori-Dankwa & K. Omane-Antwi (Eds.), *Comparative Case Studies on Entrepreneurship in Developed and Developing Countries* (pp. 1–18). Hershey, PA: IGI Global. doi:10.4018/978-1-4666-7533-9.ch001

Rahimi, R., Nadda, V., & Hamid, M. (2016). HRM Practices in Banking Sector of Pakistan: Case of National Bank of Pakistan. *International Journal of Asian Business and Information Management*, *7*(2), 25–50. doi:10.4018/IJABIM.2016040103

Raina, V. K. (2014). Overview of Mobile Payment: Technologies and Security. In F. Liébana-Cabanillas, F. Muñoz-Leiva, J. Sánchez-Fernández, & M. Martínez-Fiestas (Eds.), *Electronic Payment Systems for Competitive Advantage in E-Commerce* (pp. 186–222). Hershey, PA: IGI Global. doi:10.4018/978-1-4666-5190-6.ch011

Ramos de Luna, I., Montoro-Ríos, F., & Liébana-Cabanillas, F. J. (2014). New Perspectives on Payment Systems: Near Field Communication (NFC) Payments through Mobile Phones. In F. Liébana-Cabanillas, F. Muñoz-Leiva, J. Sánchez-Fernández, & M. Martínez-Fiestas (Eds.), *Electronic Payment Systems for Competitive Advantage in E-Commerce* (pp. 260–278). Hershey, PA: IGI Global. doi:10.4018/978-1-4666-5190-6.ch013

Rana, P., & Pandey, D. (2016). Challenges and Issues in E-Banking Services and Operations in Developing Countries. In S. Joshi & R. Joshi (Eds.), *Designing and Implementing Global Supply Chain Management* (pp. 237–281). Hershey, PA: IGI Global. doi:10.4018/978-1-4666-9720-1.ch013

Raymond, M., & Rowe, F. (2016). IS Design Considerations for an Innovative Service BPO: Insights from a Banking Case Study. *International Journal of Information Technologies and Systems Approach*, 9(2), 39–56. doi:10.4018/IJITSA.2016070103

Rossetti di Valdalbero, D., & Birnbaum, B. (2017). Towards a New Economy: Co-Creation and Open Innovation in a Trustworthy Europe. In W. Vassallo (Ed.), *Crowdfunding for Sustainable Entrepreneurship and Innovation* (pp. 20–36). Hershey, PA: IGI Global. doi:10.4018/978-1-5225-0568-6.ch002

Rouhani, S., & Savoji, S. R. (2016). A Success Assessment Model for BI Tools Implementation: An Empirical Study of Banking Industry. *International Journal of Business Intelligence Research*, 7(1), 25–44. doi:10.4018/IJBIR.2016010103

Rundshagen, V. (2014). Business Schools: Internationalization towards a New European Perspective. In A. Dima (Ed.), *Handbook of Research on Trends in European Higher Education Convergence* (pp. 124–149). Hershey, PA: IGI Global. doi:10.4018/978-1-4666-5998-8.ch007

Rusko, R., & Pekkala, J. (2014). About the Challenges to Start E-Commerce Activity in SMEs: Push-Pull Effects. In F. Musso & E. Druica (Eds.), *Handbook of Research on Retailer-Consumer Relationship Development* (pp. 490–508). Hershey, PA: IGI Global. doi:10.4018/978-1-4666-6074-8.ch026

Saiz-Alvarez, J. M. (2016). Socioeconomics of Solidarity: A Multilateral Perspective from the European Union. In J. Saiz-Álvarez (Ed.), *Handbook of Research on Social Entrepreneurship and Solidarity Economics* (pp. 192–215). Hershey, PA: IGI Global. doi:10.4018/978-1-5225-0097-1.ch011

Samoilenko, S., & Osei-Bryson, K. (2014). Investigation of Determinants of Total Factor Productivity: An Analysis of the Impact of Investments in Telecoms on Economic Growth in Productivity in the Context of Transition Economies. *International Journal of Technology Diffusion, 5*(1), 26–42. doi:10.4018/ijtd.2014010103

Sarigianni, C., Thalmann, S., & Manhart, M. (2015). Knowledge Risks of Social Media in the Financial Industry. *International Journal of Knowledge Management, 11*(4), 19–34. doi:10.4018/IJKM.2015100102

Scheepers, M. D., & Kerr, D. V. (2014). Managerial Orientations and Digital Commerce Adoption in SMEs. In P. Ordóñez de Pablos (Ed.), *International Business Strategy and Entrepreneurship: An Information Technology Perspective* (pp. 185–201). Hershey, PA: IGI Global. doi:10.4018/978-1-4666-4753-4.ch012

Sen, S., & Sen, R. L. (2014). Impact of NPAs on Bank Profitability: An Empirical Study. In N. Ray & K. Chakraborty (Eds.), *Handbook of Research on Strategic Business Infrastructure Development and Contemporary Issues in Finance* (pp. 124–134). Hershey, PA: IGI Global. doi:10.4018/978-1-4666-5154-8.ch010

Sen, S., & Sen, R. L. (2015). An Empirical Analysis of FII Movement and Currency Value in India. In N. Ray, D. Das, S. Chaudhuri, & A. Ghosh (Eds.), *Strategic Infrastructure Development for Economic Growth and Social Change* (pp. 207–217). Hershey, PA: IGI Global. doi:10.4018/978-1-4666-7470-7.ch014

Shaikh, A. A., & Karjaluoto, H. (2016). On Some Misconceptions Concerning Digital Banking and Alternative Delivery Channels. *International Journal of E-Business Research, 12*(3), 1–16. doi:10.4018/IJEBR.2016070101

Shalan, M. A. (2017). Considering Middle Circles in Mobile Cloud Computing: Ethics and Risk Governance. In K. Munir (Ed.), *Security Management in Mobile Cloud Computing* (pp. 43–72). Hershey, PA: IGI Global. doi:10.4018/978-1-5225-0602-7.ch003

Sindwani, R., & Goel, M. (2016). The Relationship between Service Quality Dimensions, Customer Satisfaction and Loyalty in Technology based Self Service Banking. *International Journal of E-Services and Mobile Applications*, 8(2), 54–70. doi:10.4018/IJESMA.2016040104

Şiriner, İ., & Shaiymbetova, K. (2016). Impacts of Global Financial Crisis and Changes in Monetary Policy of Central Banks: An Analysis of Central Bank of the Republic of Turkey (CBRT) and Bank of Israel (BOI). In M. Erdoğdu & B. Christiansen (Eds.), *Handbook of Research on Public Finance in Europe and the MENA Region* (pp. 474–504). Hershey, PA: IGI Global. doi:10.4018/978-1-5225-0053-7.ch021

Slam, M. M., & Hossain, M. E. (2015). An Investigation of Consumers' Acceptance of Mobile Banking in Bangladesh. *International Journal of Innovation in the Digital Economy, 6*(3), 16-32. doi:10.4018/ijide.2015070102

Sonmez, Y. (2015). The European Union: Another Round of Enlargement? In E. Sorhun, Ü. Hacıoğlu, & H. Dinçer (Eds.), *Regional Economic Integration and the Global Financial System* (pp. 73–87). Hershey, PA: IGI Global. doi:10.4018/978-1-4666-7308-3.ch007

Trimble, T. E. (2014). Party Rhetoric in Federal Budget Communications. In R. Hart (Ed.), *Communication and Language Analysis in the Public Sphere* (pp. 17–35). Hershey, PA: IGI Global. doi:10.4018/978-1-4666-5003-9.ch002

Tudor, C. L., & Vega, C. (2014). A Review of Textual Analysis in Economics and Finance. In R. Hart (Ed.), *Communication and Language Analysis in the Corporate World* (pp. 122–139). Hershey, PA: IGI Global. doi:10.4018/978-1-4666-4999-6.ch008

Uğurlu, M. (2016). Firm-Level Determinants of Foreign Investment and M&A Activity: Evidence from Turkey. In M. Erdoğdu & B. Christiansen (Eds.), *Handbook of Research on Comparative Economic Development Perspectives on Europe and the MENA Region* (pp. 265–292). Hershey, PA: IGI Global. doi:10.4018/978-1-4666-9548-1.ch013

Ushakov, D., & Chich-Jen, S. (2015). Global Economy Urbanization and Urban Economy Globalization: Forms, Factors, Results. In D. Ushakov (Ed.), *Urbanization and Migration as Factors Affecting Global Economic Development* (pp. 148–170). Hershey, PA: IGI Global. doi:10.4018/978-1-4666-7328-1.ch009

Uysal, Ü. E. (2017). A Brief History of City Branding in Istanbul. In A. Bayraktar & C. Uslay (Eds.), *Global Place Branding Campaigns across Cities, Regions, and Nations* (pp. 117–131). Hershey, PA: IGI Global. doi:10.4018/978-1-5225-0576-1.ch006

Valek, L. (2016). Open Ways for Time Banking Research: Project Management and Beyond. *International Journal of Human Capital and Information Technology Professionals*, 7(1), 35–47. doi:10.4018/IJHCITP.2016010103

Vardar, G., Aydoğan, B., & Acar, E. E. (2014). International Portfolio Diversification Benefits among Developed and Emerging Markets within the Context of the Recent Global Financial Crisis. In N. Ray & K. Chakraborty (Eds.), *Handbook of Research on Strategic Business Infrastructure Development and Contemporary Issues in Finance* (pp. 162–185). Hershey, PA: IGI Global. doi:10.4018/978-1-4666-5154-8.ch013

Vasudeva, S., & Singh, G. (2017). Impact of E-Core Service Quality Dimensions on Perceived Value of M-Banking in Case of Three Socio-Economic Variables. *International Journal of Technology and Human Interaction*, 13(1), 1–20. doi:10.4018/IJTHI.2017010101

Voica, M. C., & Mirela, P. (2014). Investment Development Path in the European Union in the Context of Financial Crisis. *International Journal of Sustainable Economies Management*, 3(4), 33–44. doi:10.4018/ijsem.2014100104

Wang, M., & Lin, C. (2014). Impact of Bank Operational Efficiency Using a Three-Stage DEA Model. *International Journal of Risk and Contingency Management*, 3(4), 32–50. doi:10.4018/ijrcm.2014100103

Wang, Y., Shanmugam, M., Hajli, N., & Bugshan, H. (2015). Customer Attitudes towards Internet Banking and Social Media on Internet Banking in the UK. In N. Hajli (Ed.), *Handbook of Research on Integrating Social Media into Strategic Marketing* (pp. 287–302). Hershey, PA: IGI Global. doi:10.4018/978-1-4666-8353-2.ch017

Warf, B. (2016). Digital Money in the Age of Globalization. In I. Lee (Ed.), *Encyclopedia of E-Commerce Development, Implementation, and Management* (pp. 177–183). Hershey, PA: IGI Global. doi:10.4018/978-1-4666-9787-4.ch014

Yıldırım, D. Ç., Erdoğan, S., & Gedikli, A. (2015). Fiscal Harmonization or Fiscal Union in Eurozone? In Ö. Olgu, H. Dinçer, & Ü. Hacıoğlu (Eds.), *Handbook of Research on Strategic Developments and Regulatory Practice in Global Finance* (pp. 94–104). Hershey, PA: IGI Global. doi:10.4018/978-1-4666-7288-8.ch007

Youssef, M. A. (2015). Electronic Commerce and Change in Management Accounting Practices in an Egyptian Organization. In M. Khosrow-Pour (Ed.), *Strategic E-Commerce Systems and Tools for Competing in the Digital Marketplace* (pp. 189–205). Hershey, PA: IGI Global. doi:10.4018/978-1-4666-8133-0.ch010

Zhang, L. Z. (2015). Investment Strategies for Implementing Cloud Systems in Supply Chains. In F. Soliman (Ed.), *Business Transformation and Sustainability through Cloud System Implementation* (pp. 32–43). Hershey, PA: IGI Global. doi:10.4018/978-1-4666-6445-6.ch003

About the Authors

Raghvendra Kumar has been working as Assistant Professor in Department of Computer Science and Engineering at LNCT College, Jabalpur, MP, India, and PhD Faculty of Engineering and Technology at Jodhpur National University, Jodhpur, Rajasthan. He completed his Master of Technology from KIIT University, Bhubaneswar, Odisha, and his Bachelor of Technology from SRM University, Chennai. His research interest includes graph theory, discrete mathematics, robotics, cloud computing and algorithm. He also works as reviewer, editorial and technical board member in many reputed national, international journal and conferences. He publishes research papers in international journal and conferences including IEEE and Springer regularly and supervising post graduate students in their research work.

Preeta Sharan, PhD, has a post doctorate degree from IIT Kharagpur and has 22+ years of experience in education, research, student mentorship in. She is responsible for teaching B.Tech/M.Tech. And Ph.D. Students. She has delivered lectures for the VTU e-learning program via satellite for countrywide classrooms. She is heading a strong group of R&D team working in the area of Optical Sensors, Optical Network and Quantum Dots. She has been lead investigator for several projects funded by government and international agencies such as the Naval Research Board, DRDO, IEEE USA and the Govt. of Karnataka (KVGST). Multiple patents are filed and under process. Additionally, she is also associated with the IEEE Humanitarian Technology group. She got various awards from IEEE SIGHT Asia Pacific. Recently she got the best design award at the international IEEE IMARC conference 2015. She is the Sr. member of IEEE, Execom member for IEEE Photonic Society Bangalore section, Life member for ISTE, IETE, IMAPS technical Society. She presented many papers in the various country like Singapore, China, Japan and Malaysia.

Aruna Devi is a Chartered Engineer by profession and has more than 25 years of industrial experience as an entrepreneur and 10 years as an academician at the University of Mysore (India). She is external project guide and mentor for the students of BE, MCA, MBA & M.Tech at various colleges. Her Area of expertise is in Business Analytics, Tally ERP Financial Accounting, Internet of Things, Cloud Computing and Research methodology. She has been a Life member of IEI, CSI, ISTE, MCCI, CII, NEMA, MIA & WISE. She is Hon. Secretary of CSI – Mysuru Chapter and CSI Karnataka State Student Coordinator. BOS Member, IEI Mysore Local Centre (India). She was awarded Best Women Entrepreneur - IEI (1999), Best Women Achiever-FKCCI (2013), and Successful Women Industrialists – DIC & GoK (2016). She also provides corporate IT training for industries and placement training for students at colleges.

Index

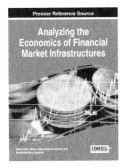

Stay Current on the Latest Emerging Research Developments

Become an IGI Global Reviewer for Authored Book Projects

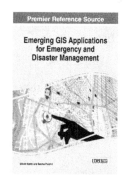

Premier Reference Source

Emerging GIS Applications for Emergency and Disaster Management

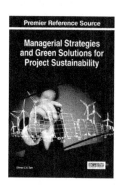

Premier Reference Source

Managerial Strategies and Green Solutions for Project Sustainability

Premier Reference Source

Comparative Approaches to Using R and Python for Statistical Data Analysis

Premier Reference Source

Solutions for High-Touch Communications in a High-Tech World

The overall success of an authored book project is dependent on quality and timely reviews.

In this competitive age of scholarly publishing, constructive and timely feedback significantly decreases the turnaround time of manuscripts from submission to acceptance, allowing the publication and discovery of progressive research at a much more expeditious rate. Several IGI Global authored book projects are currently seeking highly qualified experts in the field to fill vacancies on their respective editorial review boards:

Applications may be sent to:
development@igi-global.com

Applicants must have a doctorate (or an equivalent degree) as well as publishing and reviewing experience. Reviewers are asked to write reviews in a timely, collegial, and constructive manner. All reviewers will begin their role on an ad-hoc basis for a period of one year, and upon successful completion of this term can be considered for full editorial review board status, with the potential for a subsequent promotion to Associate Editor.

If you have a colleague that may be interested in this opportunity, we encourage you to share this information with them.

The Premier Reference for Information Science & Information Technology

100% Original Content
Contains 705 new, peer-reviewed articles with color figures covering over 80 categories in 11 subject areas

Diverse Contributions
More than 1,100 experts from 74 unique countries contributed their specialized knowledge

Easy Navigation
Includes two tables of content and a comprehensive index in each volume for the user's convenience

Highly-Cited
Embraces a complete list of references and additional reading sections to allow for further research

Included in:

InfoSci®-Books

Encyclopedia of Information Science and Technology Fourth Edition

A Comprehensive 10-Volume Set

Mehdi Khosrow-Pour, D.B.A. (Information Resources Management Association, USA)
ISBN: 978-1-5225-2255-3; © 2018; Pg: 8,104; Release Date: July 2017

For a limited time, receive the complimentary e-books for the First, Second, and Third editions with the purchase of the *Encyclopedia of Information Science and Technology, Fourth Edition* e-book.**

The **Encyclopedia of Information Science and Technology, Fourth Edition** is a 10-volume set which includes 705 original and previously unpublished research articles covering a full range of perspectives, applications, and techniques contributed by thousands of experts and researchers from around the globe. This authoritative encyclopedia is an all-encompassing, well-established reference source that is ideally designed to disseminate the most forward-thinking and diverse research findings. With critical perspectives on the impact of information science management and new technologies in modern settings, including but not limited to computer science, education, healthcare, government, engineering, business, and natural and physical sciences, it is a pivotal and relevant source of knowledge that will benefit every professional within the field of information science and technology and is an invaluable addition to every academic and corporate library.

Scan for Online Bookstore

Pricing Information

Hardcover: **$5,695**   E-Book: **$5,695***   Hardcover + E-Book: **$6,895***

Both E-Book Prices Include:
* *Encyclopedia of Information Science and Technology, First Edition E-Book*
* *Encyclopedia of Information Science and Technology, Second Edition E-Book*
* *Encyclopedia of Information Science and Technology, Third Edition E-Book*

* Purchase the Encyclopedia of Information Science and Technology, Fourth Edition e-book and receive the first, second, and third e-book editions for free. Offer is only valid with purchase of the fourth edition's e-book. Offer expires January 1, 2018.

Recommend this Title to Your Institution's Library: www.igi-global.com/books

Printed in the United States
By Bookmasters